D0940414

Springer Texts in
Electrical Engineering

Consulting Editor: John B. Thomas

John B. Thomas

Introduction to Probability

With 36 Illustrations

Springer-Verlag
New York Berlin Heidelberg Tokyo

A Dowden &
Culver Book

John B. Thomas
Department of Electrical Engineering
Princeton University
Princeton, NJ 08544
U.S.A.

Library of Congress Cataloging in Publication Data
Thomas, John Bowman, 1925–
    Introduction to probability.
    (Springer texts in electrical engineering)
    "A Dowden & Culver book."
    Bibliography: p.
    Includes index.
    1. Probability.   I. Title.   II. Series.
QA273.T425   1986   519.5   86-3768

Printed and bound by R.R. Donnelley & Sons, Harrisonburg, Virginia.
Printed in the United States of America.

9 8 7 6 5 4 3 2 1

ISBN 0-387-96319-7 Springer-Verlag New York Berlin Heidelberg Tokyo
ISBN 3-540-96319-7 Springer-Verlag Berlin Heidelberg New York Tokyo

PREFACE

This book was written for an introductory one-term course in probability. It is intended to provide the minimum background in probability that is necessary for students interested in applications to engineering and the sciences. Although it is aimed primarily at upperclassmen and beginning graduate students, the only prerequisite is the standard calculus course usually required of undergraduates in engineering and science.

Most beginning students will have some intuitive notions of the meaning of probability based on experiences involving, for example, games of chance. This book develops from these notions a set of precise and ordered concepts comprising the elementary theory of probability.

An attempt has been made to state theorems carefully, but the level of the proofs varies greatly from formal arguments to appeals to intuition. The book is in no way intended as a substitute for a rigorous mathematical treatment of probability. However, some small amount of the language of formal mathematics is used, so that the student may become better prepared (at least psychologically) either for more formal courses or for study of the literature.

Numerous examples are provided throughout the book. Many of these are of an elementary nature and are intended merely to illustrate textual material. A reasonable number of problems of varying difficulty are provided. Instructors who adopt the text for classroom use may obtain a *Solutions Manual* for all of the problems by writing to the author.

The book contains more than enough material for a one-term course. Instructors may wish to omit all or parts of Chapters 7 and 8, depending on the desired emphasis in the course and the time available. Alternatively, they may wish to cover only the first parts of both or either of these chapters. If such is the case, I would suggest Sections 7.1-7.2, 8.1-8.3, 8.5-8.6, and 8.10 or some sequential subset. Six appendices contain material that may be useful to the reader but that is outside of the mainstream of the book's development.

John B. Thomas
B-330 Engineering Quadrangle
Princeton University
Princeton, NJ   08544

TABLE OF CONTENTS

Chapter 1

INTRODUCTION AND PRELIMINARY CONCEPTS

1.1 *Random Phenomena* - Probability theory is concerned with the development of an orderly mathematical framework in which to treat *random phenomena*. These are phenomena which, under repeated observation, yield different outcomes which are not deterministically predictable. Instead the outcomes obey certain conditions of *statistical regularity* whereby the relative frequency of occurrence of the possible outcomes is approximately predictable. A classical example of such a random phenomenon is the tossing of an unbiased coin. In this case, two outcomes are possible - heads (H) and tails (T). Although it is not possible to predict whether a particular toss will yield a head or a tail, it can usually be expected that, in a long sequence of tosses, approximately half the outcomes will be "heads" and half will be "tails". On such considerations we say that the "probability" of a head is one-half, as is the "probability" of a tail. However, as we shall see later, relative frequency of occurrence is more useful mathematically as an interpretation of probability rather than as a definition.

Nevertheless, inherent to the consideration of random phenomena is a repeated experiment with a set of *possible outcomes* or *events*. Associated with each of these events is a real number called the *probability of the event*. This probability is related in some manner to the expected relative frequency of occurrence of the event in a long sequence of experiments. It seems intuitively reasonable that each probability should lie between zero and unity and that the sum of the probabilities of the events for a particular experiment should be unity.

This study of probability theory will begin with a consideration of the relations among the events (possible outcomes) associated with a random experiment. The language and elementary ideas of *set theory* will be used to describe these relations, and this topic will be treated in the next section of this chapter. As will become clear as we proceed, the collection of events will correspond to the *universal set* and will be called the *sample space* while the events themselves will correspond to *subsets* of the sample space. Each event will be composed of *elementary events* or *elements* or *points* of the sample space. These points are not further decomposable and comprise the ultimate building blocks of the sample space.

1.2 *Elements of Set Theory* - We begin with the undefined notion of a *set*, which will be taken to mean *a collection, an aggregate, a class, or a family of any objects whatsoever*. Each of the objects which make up the set is called an *element*, or *member*, or *point* of the set.

It will be convenient to denote sets by upper case letters and elements of sets by lower case letters. For example, the letter A will denote a given set and

$$a \in A$$

will indicate that a is an element of the set A. It will be said that a "belongs" to A. If the element a does not belong to the set A, we write

$$a \notin A$$

It is conventional to specify a set in one of two ways:

(1) by listing the elements of the set between braces, or

(2) by listing some property common to all the elements of the set.

For example, let I be the set of all integers and consider the set B with elements 1, 2, and 3. We write:

$$B = \{1, 2, 3\}$$

or

$$B = \{b \in I \mid 0 < b < 4\}$$

The last expression is read "B is the set of all integers b which are greater than zero and less than four."

It should be noted that the order of the elements is immaterial and that no element can appear more than once. Thus the set $\{1,2,3\}$ and the set $\{3,1,2\}$ are the same, and the collection $\{1,1,2,3,1\}$ is a set only if the repeated elements are deleted.

We shall use equality in the sense of identity. Two sets A and B are equal if, and only if, they contain exactly the same elements. In this case we write

$$A = B$$

If the elements in the two sets A and B are not the same, we write

$$A \neq B$$

and say that A and B are *distinct*.

If every element of B is also an element of A, we say *that B is contained in A* or *B is a subset of A* and write

$$B \subseteq A$$

$$A \supseteq B$$

and say that A contains B. If $B \subseteq A$ and at least one element of A is not in B, we call B a *proper subset* of A and say that A *properly contains B*

$$A \supset B$$

or that B *is properly contained in A*

$$B \subset A$$

Since it may be necessary to consider a particular set which will turn out to have no elements, it is convenient to define the *void*, or *empty*, or *null* set ϕ as the set which contains no elements.

In most discussions of sets a particular class of objects will exist which comprise an all-embracing set. The sets with which we concern ourselves will then be sets of elements from this fixed set. This fixed set is known as the *universal set U* or the *sample space S*. The two terms will be used interchangeably. Then, for every set A, it follows that

$$\phi \subseteq A \subseteq S$$

and, for every element a, that

$$a \in S \text{ and } a \notin \phi$$

It is sometimes convenient to attach a geometric significance to the subsets of a given sample space. By this convention areas are associated with sets and points with elements. The sample space S is often indicated by a rectangle and subsets of the space by areas within the rectangle. Figure 1.1 illustrates the

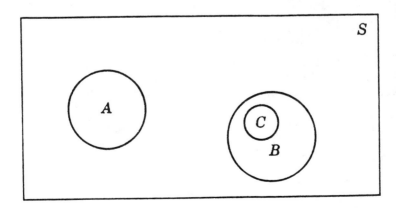

Fig. 1.1 - A Venn diagram

technique. Here A, B, and C are subsets of the sample space S. In addition, C is a proper subset of B and we could write

$$C \subset B$$

These diagrams are called *Venn diagrams*.

At this point it is convenient to consider some of the ways in which sets can be combined with one another and with some of the properties of such combinations.

The *intersection* of two sets A and B is denoted by

$$A \cap B \quad or \quad AB$$

and is the set of all elements common to both A and B; that is,

$$A \cap B = AB = \{x \mid x \in A \text{ and } x \in B\}$$

The intersection AB is sometimes called the *event A and B*. The *union* of two sets A and B is denoted by $A \cup B$ and is the set of all elements in A or B or both; that is,

$$A \cup B = \{x \mid x \in A \quad or \quad x \in B\}$$

The union $A \cup B$ is sometimes called* the *event A or B*. The Venn diagram of Figure 1.2 shows the union and intersection of A and B for a typical situation.

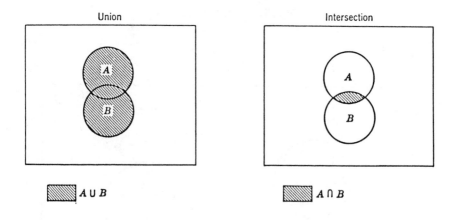

Fig. 1.2 - Unions and intersections of sets

---

*Note that "or" is used here in the sense of "either..or ..or both".

The *complement* of the set A is denoted by A^c ($or A'$) and is the set of all elements not in A ; that is,

$$A^c = \{x \mid x \notin A\}$$

Some useful relationships involving the complement are

$$(A^c)^c = A$$

$$\phi^c = S \qquad S^c = \phi$$

$$A \cup A^c = S \qquad A \cap A^c = \phi$$

$$(A \cup B)^c = A^c \cap B^c \qquad (A \cap B)^c = A^c \cup B^c$$

If two sets A and B have no elements in common they are said to be *disjoint* or *mutually exclusive* and

$$AB = \phi$$

It has already been emphasized that the elements of a set can be anything whatsoever; in particular, they can be sets themselves. To avoid confusion, it is conventional to call

a *set of sets*   a *class of sets*

and

a *set of classes*   a *family (of classes)*.

A set will be called a *finite set* if it is empty or contains exactly n elements, for n a positive integer; otherwise, the set will be called an *infinite set*.

Based on the previous definitions, all subsets of the universal set S form an algebraic system, for which the following theorems hold. For completeness, some relationships are repeated that have been given previously.

$$A \cup \phi = A \qquad A \cup S = S$$
$$A \cap \phi = \phi \qquad A \cap S = A$$
$$A \cup A = A \qquad S \cup \phi = S$$
$$A \cap A = A \qquad S \cap \phi = \phi$$

If $A \subset B$, then $(A \cup B) = B$ and $(A \cap B) = A$.

For all A,B; $A \subseteq (A \cup B)$ and $(A \cap B) \subseteq A$.

If $B \subset C$, then $(A \cap B) \subset (A \cap C)$.

If $A \subset C$, and $B \subset C$, then $(A \cup B) \subset C$.

If $A \subset C$ and $B \subset D$, then $(A \cup B) \subset (C \cup D)$.

The commutative law

$$A \cap B = B \cap A$$
$$A \cup B = B \cup A$$

The distributive law

$$A \cap (B \cup C) = (A \cap B) \cup (A \cap C)$$
$$A \cup (B \cap C) = (A \cup B) \cap (A \cup C)$$

The associative law

$$(A \cup B) \cup C = A \cup (B \cup C) = A \cup B \cup C$$
$$(A \cap B) \cap C = A \cap (B \cap C) = A \cap B \cap C$$

The ideas of union and intersection may be extended in a rather obvious fashion to n sets. If $A_1, A_2,...,A_n$ are subsets of S, the union of $A_1, A_2,...,A_n$ is the set of all elements which belong to at least one of the sets $A_1, A_2,...,A_n$ and may be denoted by

$$A_1 \cup A_2 \cup \cdots \cup A_n \text{ or } \bigcup_{i=1}^{n} A_i$$

The intersection of $A_1, A_2,...,A_n$ is the set of all elements common to $A_1, A_2,...A_n$ and is written as

$$A_1 \cap A_2 \cap \cdots \cap A_n \text{ or } \bigcap_{i=1}^{n} A_i$$

Similarly, let ξ be a family of subsets of S. The set of all $x \in S$ such that $x \in A$ *for at least one* $A \in \xi$ can be written as

$$\bigcup_{A \in \xi} A = \{x \in S \mid x \in A \text{ for } some \ A \in \xi\}$$

and can be called the union of the sets of ξ. In the same way, the set of all $x \in S$ such that $x \in A$ *for every* $A \in \xi$ is the intersection of the sets of ξ and is written

$$\bigcap_{A \in \xi} A = \{x \in S \mid x \in A \text{ for } all \ A \in \xi\}$$

An infinite set is said to be *countable* or *denumerable* if its elements can be indexed by the natural numbers (positive integers); that is, placed in one-to-one correspondence with them. Otherwise an infinite set is said to be *uncountable* or *non-denumerable*. As an example, the set of natural numbers is a countable set; the set of all real numbers is not.

The relationships involving the complement may be extended to the countable case to yield the following set of useful rules know as *DeMorgan's rules:*

$$\left(\bigcup_{i=1}^{\infty} A_i\right)^c = \bigcup_{i=1}^{\infty} A_i^c \quad , \quad \left(\bigcap_{i=1}^{\infty} A_i\right)^c = \bigcup_{i=1}^{\infty} A_i^c$$

$$\left(\bigcup_{i=1}^{\infty} A_i^c\right)^c = \bigcap_{i=1}^{\infty} A_i^c \quad , \quad \left(\bigcap_{i=1}^{\infty} A_i^c\right)^c = \bigcup_{i=1}^{\infty} A_i$$

We proceed now to develop briefly a concept of probability based on the idea of *equally likely* outcomes. Later, it will be shown that the relationships developed can be obtained more rigorously from an axiomatic approach.

1.3 A Classical Concept of Probability - Much of the historical foundation of probability arose from a consideration of games of chance where the possible outcomes could be reduced to elementary events which were *equally likely*. The term "equally likely" is taken to have the following meaning. If the possible outcomes of a random experiment are the M (elementary) events $x_1, x_2, ..., x_M$, then the probability of each event is

$$P(x_1) = P(x_2) = \cdots = P(x_M) = \frac{1}{M} \qquad (1.3\text{-}1)$$

A set of these elementary events is said to be *favorable* to an event E if the event E occurs when, and only when, one of the members of the set occurs. Then the probability of the event E is

$$P(E) = \frac{m}{M}$$

where m is the total number of elements in the set favorable to E. Thus the calculation of the probability of any event reduces to finding two numbers: 1) the total number M of possible outcomes in the experiment and 2) the number m of outcomes favorable to the particular event.

Example 1.1

An unbiased coin is tossed three times: Find the probability that *at least* two of the tosses yield "heads". Find the probability that *exactly* one head is thrown.

The set of all possible outcomes consists of the following eight members:

$$HHH \quad HHT \quad HTT \quad TTT$$
$$HTH \quad THT$$
$$THH \quad TTH$$

Of these, four satisfy the condition of containing at least two heads. The required probability is

$$P(at\ least\ two\ heads) = \frac{4}{8} = \frac{1}{2}$$

There are three outcomes favorable to the event "exactly one head"; hence

$$P(\textit{exactly one head}) = \frac{3}{8}$$

This classical concept of probability suffers from an obvious shortcoming. It depends completely on *a priori* analysis of the random experiment. All possible outcomes must be found and a clear relation must be established between these outcomes and a set of equally likely events comprising the elementary events of the sample space. Nevertheless, some of the tools necessary to apply this concept will be treated in the following section. These ideas will be useful, also, for the remainder of the text.

1.4 *Elements of Combinatorial Analysis* - Whenever equal probabilities are assigned to the elements of a finite sample space, the computation of probabilities of events reduces to the counting of the elements comprising the events. For such purposes, some of the introductory concepts of combinatorial analysis will prove immensely useful.

(a) *Pairs and Multiplets*

A *pair* (a_j, b_k) is an *unordered set of two elements* a_j and b_k. Thus (a_j, b_k) and (b_k, a_j) will be considered the same pair. With m elements $a_1, a_2, ..., a_m$ and n elements $b_1, b_2, ..., b_n$ it is possible to form mn pairs (a_j, b_k) containing one element from each group since there are n of the b_k' s that can be paired with each of the m a_j' s. Similarly a *multiplet* of size r is an *unordered set of r elements*. Given n_1 elements $a_1, a_2, ..., a_{n_1}$; n_2 elements $b_1, b_2, ..., b_{n_2}$; etc.; up to n_r elements $x_1, x_2, ..., x_{n_r}$, it is possible to form $n_1 n_2 ... n_r$ multiplets $(a_j, b_k, ..., x_l)$ containing one element from each group.

(b) *Sampling*

A *sample* of size r is an *ordered set of r elements* obtained in some way from a universal set of n elements. Two possible policies to follow in obtaining samples are 1) to *sample with replacement* of the elements drawn and 2) to *sample without replacement*.

Consider first the case of sampling with replacement. There are n possible elements any one of which can be chosen on the first drawing to form the first element in the sample. Since the chosen element is replaced prior to the second drawing, there are also n possible elements available to be chosen as the second element in the sample. Thus, after r drawings, the sample (of length r) could be any one of

$$n \ . \ n \ . \ n \ ... \ n = n^r \qquad (1.4\text{-}1)$$

distinct possible samples.

In sampling without replacement it is apparent that $r \leq n$. After the first drawing, there are only $n-1$ elements available from which to choose and, after r drawings, only $n-r$. Thus the number of possible samples of length r in this case is only

$$n(n-1)(n-2) \; \ldots \ldots \; (n-r+1) = \frac{n!}{(n-r)!} \qquad (1.4\text{-}2)$$

We note in passing that

$$\frac{n!}{(n-r)!} \leq n! \leq n^n \qquad (1.4\text{-}3)$$

where r and n are non-negative integers and $r \leq n$. An (ordered) sample taken without replacement (so that the same element does not occur twice) is often called a *permutation*. The symbol p_r^n is sometimes used to indicate the number of permutations of n things taken r at a time:

$$p_r^n = \frac{n!}{(n-r)!} \qquad (1.4\text{-}4)$$

Suppose we sample without replacement to obtain a sample of size n from a set of n elements. Then the number of such samples is just $n!$ since

$$\frac{n!}{(n-n)!} = n! \qquad (1.4\text{-}5)$$

Furthermore this is the number of ways of *ordering* the n elements since each sample is a unique ordering and the $n!$ possible samples exhaust the possibilities.

(c) *Combinations*

A *combination* is a set of elements without repetition and without regard to order. Thus the set $\{1,2\}$ and the set $\{2,1\}$ are different permutations but only one combination. Out of n elements it has already been shown that the possible permutations of size r are

$$\frac{n!}{(n-r)!}$$

We have seen that r elements can be order in $r!$ ways so that there are $r!$ permutations containing any particular set of r elements. Therefore the number of possible combinations of n elements taken r at a time must be

$$\frac{n!}{(n-r)!} \div r! = \frac{n!}{(n-r)!\,r!} \qquad (1.4\text{-}6)$$

This quantity is frequently denoted by C_r^n or by $\begin{pmatrix} n \\ r \end{pmatrix}$ and is often called the *binomial coefficient* since it occurs in the binomial expansion

$$(a+b)^n = \sum_{r=0}^{n} \left(\begin{array}{c} n \\ r \end{array} \right) a^{n-r} b^r \qquad (1.4\text{-}7)$$

It follows directly from Eq. (1.4-6) that

$$\left(\begin{array}{c} n \\ r \end{array} \right) = \left(\begin{array}{c} n \\ n-r \end{array} \right) \qquad (1.4\text{-}8)$$

and that

$$\left(\begin{array}{c} n \\ n \end{array} \right) = 1, \quad \left(\begin{array}{c} n \\ 0 \end{array} \right) = 1 \qquad (1.4\text{-}9)$$

Example 1.2

What is the probability that a bridge hand will contain all thirteen card values: that is, A,K,Q,.....,3,2?

Since there are 52 cards in a deck of bridge cards and a hand contains 13 cards, the number of possible bridge hands is given by

$$\left(\begin{array}{c} 52 \\ 13 \end{array} \right)$$

Since the deck contains four cards of each of the thirteen values, the number of possible bridge hands with all thirteen card values is

$$4^{13}$$

Thus the required probability is

$$4^{13} \div \left(\begin{array}{c} 52 \\ 13 \end{array} \right) \approx 0.0001057$$

(d) *Multinomial Coefficients*

Let $r_1, r_2, ..., r_k$ be non-negative integers such that

$$r_1 + r_2 + \cdots + r_k = n$$

It will now be shown that the number of ways in which n objects can be divided into k groups, the first containing r_1 objects, the second r_2, etc. is given by the multinomial coefficient

$$\left(\begin{array}{c} n \\ r_1,...,r_k \end{array} \right) = \frac{n!}{r_1!...r_k!} \qquad (1.4\text{-}10)$$

For the case where $k=2$, the multinomial coefficient reduces to the binomial coefficient $\begin{pmatrix} n \\ r_1 \end{pmatrix}$ since $r_2 = n - r_1$. In the general case, we begin by selecting the first group of size r_1 from the set of n objects. There will be $\begin{pmatrix} n \\ r_1 \end{pmatrix}$ combinations possible. Next we select the second group of size r_2 from the remaining $n - r_1$ objects. There will be $\begin{pmatrix} n - r_1 \\ r_2 \end{pmatrix}$ combinations possible. We continue in this manner until the $(k-1)$ group has been chosen. After this choice, there are only r_k objects left and they must comprise the k-th group. Thus the total number of ways in which the n objects can be divided into the k groups is

$$\begin{pmatrix} n \\ r_1 \end{pmatrix} \begin{pmatrix} n - r_1 \\ r_2 \end{pmatrix} \cdots \begin{pmatrix} n - r_1 - r_2 - \ldots - r_{k-2} \\ r_{k-1} \end{pmatrix}$$

This product may be written explicitly as

$$\frac{n!}{r_1!(n-r_1)!} \frac{(n-r_1)!}{r_2!(n-r_1-r_2)!} \cdots \frac{(n-r_1-r_2-\ldots-r_{k-2})!}{(r_{k-1})!r_k!}$$

On canceling similar terms in the numerator and denominator it is clear that this last expression is the multinomial coefficient originally defined.

Example 1.3

Twelve dice are thrown. What is the probability that each face occurs twice; that is, that there are two 1's, two 2's, etc?

The total number of possible outcomes is 6^{12} since each die has six distinct faces. Each face value can occur twice in the twelve throws in as many ways as twelve dice can be arranged in six groups of two each. In terms of the multinomial coefficient, $n=12$ and $r_1 = r_2 = \cdots = r_6 = 2$. Thus the total number of favorable outcomes is

$$\frac{12!}{2\times2\times2\times2\times2\times2} = \frac{12!}{2^6}$$

and the probability desired is

$$p = \frac{12!}{2^6 6^{12}} \approx 0.003438$$

(e) *The Hypergeometric Distribution*

Consider a set of n elements. Suppose that

n_1 of these elements have a particular attribute while the remainder $n_2 = n - n_1$ have some other attribute. A group of r elements is chosen at random without replacement and without regard to order. It is desired to find the probability $q(k)$ that, of the group chosen, exactly k elements have the first attribute. We proceed as follows:

The chosen group has k elements with the first attribute and $r-k$ elements with the second. The subgroup with the first attribute can be chosen in $\binom{n_1}{k}$ ways and the subgroup with the second attribute in $\binom{n-n_1}{r-k}$ ways. Therefore the total number of favorable outcomes is

$$\binom{n_1}{k}\binom{n-n_1}{r-k}$$

The total number of possible outcomes is just the number of combinations of n things taken r at a time and is given by $\binom{n}{r}$. Thus the probability $q(k)$ is

$$q(k) = \frac{\binom{n_1}{k}\binom{n-n_1}{r-k}}{\binom{n}{r}} \tag{1.4-11}$$

The function $q(k)$ is called the *hypergeometric distribution function*.

Example 1.4

Find the probability that, among r cards drawn at random from a bridge deck, there are exactly four aces.

In this case $n = 52$, $n_1 = 4$, and $k = 4$. We have

$$q(4) = \frac{\binom{4}{4}\binom{52-4}{r-4}}{\binom{52}{r}} = \frac{\binom{48}{r-4}}{\binom{52}{r}} \ , \quad r = 4,5,...,52$$

(f) *Stirling's Formulas*

In the previous sections the factorial expression

$$n! = n(n-1)(n-2)....(3)(2)(1)$$

has occurred frequently. For small n, this expression is easily evaluated, but the computation may become prohibitively difficult as n increases. Furthermore a number of limiting expressions will be encountered later where it will be convenient to have available approximations to $n!$. An approximation for $n!$ which converges to $n!$ in the limit is [Feller (1968); pp. 52-54]

$$n! \approx (2\pi)^{1/2} n^{n+1/2} e^{-n} \tag{1.4-12}$$

This expression is frequently called *Stirling's first approximation* and underestimates $n!$. A *second approximation* which overestimates $n!$ is

$$n! \approx (2\pi)^{1/2} n^{n+1/2} e^{-n+1/12n} \tag{1.4-13}$$

Even for small n, the *first approximation* is remarkably accurate.

1.5 *The Axiomatic Foundation of Probability Theory* - The theory of probability can be developed formally from a set of axioms. The approach to be used here was due originally to Kolmogorov (1956). The language used will be that of the set theory discussed in Section 1.2. The *elements* of the *sample space S* will be called *outcomes* or *elementary events* or *points* of a random experiment and will be denoted as before by lower case letters $a, b, ...$ or sometimes by ω. Certain subsets of the sample space will be called *events* and will be denoted by upper case letters $A, B, ...$; included will be the *certain event S* and the *impossible event ϕ*. The sample space may be *finite, denumerably infinite,* or *non-denumerably infinite.* Of these the finite sample space is the simplest and will be discussed first.

1.6 *Finite Sample Spaces* - A finite sample space has been defined previously as a set which is either empty or contains exactly n elements where n is a positive integer. A finite sample space S will be called a *probability space* if, for every event $A \in S$, there is defined a real number $P(A)$ with the properties:

Axiom 1.

$$P(A) \geq 0. \tag{1.6-1}$$

Axiom 2.

$$P(S) = 1 \tag{1.6-2}$$

Axiom 3.

If A_1 and A_2 are two disjoint events in S (that is, if $A_1 \cap A_2 = \phi$), then

$$P(A_1 \cup A_2) = P(A_1) + P(A_2) \tag{1.6-3}$$

The function $P(A)$ is called the *probability of the event A*. Note that an immediate consequence of Axiom 3 is

Axiom 3(a).

Consider the mutually disjoint events $A_1, A_2, \ldots, A_m$; then

$$P(A_1 \cup A_2 \cup \cdots \cup A_m) =$$

(1.6-4)

$$P(A_1) + (P(A_2) + \cdots + P(A_m)$$

Proof: For $m=2$, this is a restatement of Axiom 3. Let $m=3$ and define the event A by $A = A_1 \cup A_2$. Since the A_i are mutually disjoint, as discussed in Section 1.2, $A \cap A_3 = \phi$ and $A_1 \cap A_2 = \phi$ so that

$$P(A \cup A_3) = P(A) + P(A_3) = P(A_1) + P(A_2) + P(A_3)$$

In the same way define the event B by $B = A_1 \cup A_2 \cup \cdots \cup A_{m-1}$ where $B \cap A_m = \phi$ and $A_i \cap A_j = \phi$ for $1 \leq i, j \leq m-1$, $i \neq j$. Again, by Axiom 3

$$P(B \cup A_m) = P(B) + P(A_m) = \sum_{i=1}^{m} P(A_i) \qquad \bullet$$

For infinite sample spaces some modifications of the axioms are necessary and some additional concepts of set theory are required.

1.7 *Fields, σ–Fields, and Infinite Sample Spaces* - As mentioned in Section 1.2, a set, whose elements are sets, will be called a *class* of sets. These classes will usually be denoted by bold-faced letters such as **A** or **B**. When the set operations performed on the sets in a class **A** give as a result sets which also belong to this class, then the class **A** is said to be *closed* under those set operations. A *field (or algebra)* **F** is a class of sets which is closed under complementation and finite intersections and unions. Thus a field **F** is a nonempty class of sets of S such that

(1) If $A \in \mathbf{F}$ and $B \in \mathbf{F}$, then $A \cup B \in \mathbf{F}$

(2) If $A \in \mathbf{F}$, then $A^c \in \mathbf{F}$

These two properties are sufficient to define a field. A number of other properties follow immediately. For example:

(3) If $A \in \mathbf{F}$ and $B \in \mathbf{F}$, then $A \cap B = AB \in \mathbf{F}$

Proof: From (1) and (2) it follows that $A^c \in \mathbf{F}, B^c \in \mathbf{F}$, and $A^c \cup B^c \in \mathbf{F}$; therefore $(A^c \cup B^c)^c = AB \in \mathbf{F}$. $\bullet$

(4) $S \in \mathbf{F}$

Proof: Since the class is nonempty, at least one set $A \in \mathbf{F}$. Hence $A^c \in \mathbf{F}$ and $A \cup A^c = S \in \mathbf{F}$. $\bullet$

(5) $\phi \in \mathbf{F}$

Proof: Since $S \in \mathbf{F}$, its complement $S^c = \phi \in \mathbf{F}$. •

(6) If $A \in \mathbf{F}$ and $B \in \mathbf{F}$, then $A - B \in \mathbf{F}$

Proof: It has been shown that $B^c \in \mathbf{F}$, and, hence, $AB^c = A - B \in \mathbf{F}$. •

A field is sometimes called an *additive* class of sets. It is clear from the properties just developed that all of the set operations can be performed any (finite) number of times on the members of the field $\mathbf{F}$ without obtaining a set not in $\mathbf{F}$.

Example 1.5

Consider the set $S = 1,2$. Two possible fields $\mathbf{F}_1$ and $\mathbf{F}_2$ are

$$\mathbf{F}_1 = \{\phi, \{1,2\}\}$$
$$\mathbf{F}_2 = \{\phi, \{1\}, \{2\}, \{1,2\}\}$$

Any field $\mathbf{B}$ on S is called a *completely additive class* or a *σ-field* (sigma field)* on S if it is closed under *denumerable* intersections and unions. Thus, a *σ-field* $\mathbf{B}$ is a *field on S* such that

(1) If $A_1, A_2, ..., A_n, ... \in \mathbf{B}$, then $\bigcup_{i=1}^{\infty} A_i \in \mathbf{B}$

It follows that

(2) If $A_1, A_2, \ldots, A_n, ... \in \mathbf{B}$, then $\bigcap_{i=1}^{\infty} A_i \in \mathbf{B}$

Proof: If $A_n \in \mathbf{B}$, then $A_n^c \in \mathbf{B}$ and $\bigcup_{n=1}^{\infty} A_n^c \in \mathbf{B}$. In the same way the complement $(\bigcup_{n=1}^{\infty} A_n^c)^c \in \mathbf{B}$. It follows from one of DeMorgan's rules (see Section 1.2) that

$$(\bigcup_{n=1}^{\infty} A_n^c)^c = \bigcap_{n=1}^{\infty} A_n \in \mathbf{B} \quad •$$

The elements of the *σ-field* will be the *events* of the random experiments. In a given sample space S it is generally possible to define many *σ-fields*. To distinguish between members of several *σ-fields*, the members of a given *σ-field* $\mathbf{B}$ are usually called $\mathbf{B}$ *-measurable sets*.

---

*The completely additive class of sets on the real line containing all intervals is usually called the *Borel class* or *Borel field*. A more complete discussion of the construction of this field is given in Cramer (1946) and Dubes (1968).

A *probability space* or *field of probability* can now be defined as the triple** $(S,\mathbf{B},P)$ where S is a sample space (finite or infinite), $\mathbf{B}$ is a $\sigma{-}field$ on S, and P is a real-valued *set function*** such that, for every event $A \in \mathbf{B}$,

Axiom 1.

$$P(A) \geq 0 \qquad\qquad (1.7\text{-}1)$$

Axiom 2.

$$P(S) = 1 \qquad\qquad (1.7\text{-}2)$$

Axiom 3.

If $A_1, A_2,...,A_n,...$ is any countable sequence of mutually disjoint events in $\mathbf{B}$, then

$$P(A_1 \cup A_2 \cup ...\cup A_n \cup ...) = \sum_{i=1}^{\infty} P(A_i) \qquad (1.7\text{-}3)$$

The function $P(A)$ is called the *probability* of the event A. If S is a finite space and if $\mathbf{B}$ is chosen to be all of the events in S, then these axioms are equivalent to those used in Section 2.2 to define a finite probability space. In all future discussions of events and their associated probabilities, it should be kept in mind that there is an underlying probability space $(S,\mathbf{B},P)$ consisting of a sample space, a $\sigma{-}field$, and a probability measure whether these are mentioned specifically or not.

Example 1.6

Let the sample space S consist of the single point b so that

$$S = \{b\}$$

A $\sigma{-}field$ $\mathbf{B}$ on S is given by

$$\mathbf{B} = \{S, \phi\}$$

and a probability set function is

$$P(S) = 1 \quad , \quad P(\phi) = 0$$

** Many treatments of probability use the notation Ω instead of S for the sample space or *fundamental* probability set. The points or elementary events of Ω are then usually denoted by ω.
***A *set function* is a function whose domain of definition is a class of sets; i.e., a rule exists which relates some object or group of objects to each member of the class. In this case P is a real-valued function which assigns to every $A \in \mathbf{B}$ a number P(A).

The probability $P(\cdot)$ possesses certain properties that are an immediate consequence of its definition.

Corollary 1.

$$P(\phi) = 0 \qquad (1.7\text{-}4)$$

Proof: Since $S = S \cup \phi$ and $\phi = S\phi$, it follows from Axiom 3 that $P(S) = 1 = P(S) + P(\phi) = 1 + P(\phi)$; thus $P(\phi) = 0$. •

Corollary 2.

$$P(A^c) = 1 - P(A) \qquad (1.7\text{-}5)$$

Proof: The sample space S can be written as $S = A \cup A^c$ and $AA^c = \phi$; it is clear then that $P(A \cup A^c) = P(A) + P(A^c) = 1$ and Corollary 2 follows. Note that Corollary 1 results if $A = S$. •

Corollary 3. For any finite set $E_1, E_2, ..., E_n$ of disjoint events,

$$P(E_1 \cup E_2 \cup \cdots \cup E_n) = \sum_{i=1}^{n} P(E_i) \qquad (1.7\text{-}6)$$

Proof: Let the null set $\phi = E_{n+1} = E_{n+2} = \cdots$. It follows from Axiom 3 and Corollary 1 that

$$P(E_1 \cup \cdots \cup E_n) = P(\bigcup_{i=1}^{\infty} E_i) = \sum_{i=1}^{\infty} P(E_i) = \sum_{i=1}^{n} P(E_i) \quad •$$

Corollary 4. Let E and F be events and let $E \subseteq F$; then

$$P(E) \le P(F) \qquad (1.7\text{-}7)$$

Proof: The event F can be written as $E \cup E^c F$. Now E and $E^c F$ are disjoint; that is, $E \cap E^c F = \phi$. Then, by Axiom 1 and Corollary 3, we have

$$P(F) = P(E) + P(E^c F) \ge P(E) \quad •$$

Corollary 5. For every event $E \in \mathbf{B}$,

$$P(E) \le 1 \qquad (1.7\text{-}8)$$

Proof: The relation $E \in \mathbf{B}$ implies $E \subseteq S$. Then, by Axiom 2 and Corollary 4, it follows that $P(E) \le P(S) = 1$. •

Corollary 6. (Boole's Inequality). Consider any sequence of events $A_1, A_2, ..., A_n, ...$ not necessarily disjoint; then

$$P(\bigcup_{i=1}^{\infty} A_i) \le \sum_{i=1}^{\infty} P(A_i) \qquad (1.7\text{-}9)$$

Proof: The proof follows from the equality

$$\bigcup_{i=1}^{\infty} A_i = A_1 \cup (A_2 - A_1) \cup [A_3 - (A_1 \cup A_2)] \cup \cdots \qquad (1.7\text{-}10)$$

Note that the right side of this expression consists of a union of disjoint events; hence

$$P\left(\bigcup_{i=1}^{\infty} A_i\right) =$$

(1.7-11)

$$P(A_1) + P(A_2 - A_1) + P[A_3 - (A_1 \cup A_2)] + \cdots$$

Since $(A_2 - A_1) \subseteq A_2$, $[A_3 - (A_1 \cup A_2)] \subseteq A_3$, etc., it follows that $P(A_2 - A_1) \leq P(A_2)$, $P[A_3 - (A_1 \cup A_2)] \leq P(A_3)$, etc., and Eq. (1.7-9) follows immediately. •

1.8 *Conditional Probability and Independence* - Consider two events A and B such that $A \in \mathbf{B}$ and $B \in \mathbf{B}$. Assume that $P(B) > 0$. The *conditional probability* of A given B will be denoted by $P(A/B)$ and will be defined as

$$P(A/B) = \frac{P(AB)}{P(B)}$$

(1.8-1)

For fixed $B \in \mathbf{B}$ with $P(B) > 0$, the function $P(\cdot/B)$ on $\mathbf{B}$ is a probability. In other words, as a function of $A \in \mathbf{B}$, the function $P(A/B)$ satisfies the three axioms used to define a probability; that is, for every event $A \in \mathbf{B}$

Axiom 1(a)

$$P(A/B) \geq 0$$

(1.8-2a)

Axiom 2(a)

$$P(S/B) = 1$$

(1.8-2b)

Axiom 3(a)

If $A_1, A_2, ..., A_n, ..., ...$ is any countable sequence of mutually disjoint events in $\mathbf{B}$, then

$$P\left(\bigcup_{i=1}^{\infty} A_i/B\right) = \sum_{i=1}^{\infty} P(A_i/B)$$

(1.8-3)

Proof of Axiom 1(a): The probability $P(AB)$ satisfies the inequality $0 \leq P(AB) \leq 1$ while $P(B)$ satisfies $0 < P(B) \leq 1$. Axiom 1(a) follows from Eq. (1.8-1). •

Proof of Axiom 2(a): An arbitrary event $B \in \mathbf{B}$ satisfies the identity $B = SB$; hence, from Eq. (1.8-1), we have

$$P(S/B) = \frac{P(SB)}{P(B)} = \frac{P(B)}{P(B)} = 1 \qquad •$$

Proof of Axiom 3(a): Since the events $A_1, A_2, ..., A_n, ...$ are mutually disjoint, the same is true of the sequence $A_1 B, A_2 B, ..., A_n B, ...$ Consider the relationships

$$P(\bigcup_{i=1}^{\infty} A_i / B) = \frac{P(\bigcup_{i=1}^{\infty} A_i B)}{P(B)} = \frac{\sum_{i=1}^{\infty} P(A_i B)}{P(B)} = \sum_{i=1}^{\infty} P(A_i / B) \qquad \bullet$$

Example 1.7

Let B be the arbitrary event shown in Fig. 1.3 with $P(B) > 0$. Let S be a finite space with all elements equally likely. Assume that $|S|$ is the number of elements in S, that $|B|$ is the number of elements in B, and that $|A|$ is the number of elements in A. We have

$$P(B) = \frac{|B|}{|S|}, \quad P(A) = \frac{|A|}{|S|}, \quad \text{and} P(AB) = \frac{|AB|}{|S|}$$

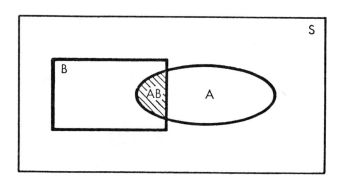

Fig. 1.3 - Conditional Probability Relationships

However, it follows from Eq. (1.8.1) that

$$P(A/B) = \frac{P(AB)}{P(B)} = \frac{|AB|}{|S|} \frac{|S|}{|B|} = \frac{|AB|}{|B|}$$

Conditioning the probability of Event A on the occurrence of Event B has the effect of renormalizing to a reduced sample space B instead of the original sample space S.

Consider now the sequence of disjoint events $\{B_n\}$. The sequence may be finite or countable infinite. Assume that

$$P(\bigcup_n B_n) = \sum_n P(B_n) = 1 \qquad (1.8\text{-}4)$$

The event A is an arbitrary event in $\mathbf{B}$ and $P(B_n) > 0$ for every n. The probability of event A may be written as

$$P(A) = P[A \cap (\bigcup_n B_n)] + P[A \cap (\bigcup_n B_n)^c] \qquad (1.8\text{-}5)$$

But the probability $P(\bigcup_n B_n)^c$ is zero by Eq. (1.8-4). Consequently, this last equation becomes

$$P(A) = P[A \cap (\bigcup_n B_n)] = P(\bigcup_n AB_n) = \sum_n P(AB_n) \qquad (1.8\text{-}6)$$

or, finally,

$$P(A) = \sum_n P(AB_n) = \sum_n P(A/B_n)P(B_n) \qquad (1.8\text{-}7)$$

This last expression is frequently called the *Theorem of Total Probabilities.*

Consider the events A and B, both assumed to have probabilities different from zero. If

$$P(A/B) = P(A) , \qquad (1.8\text{-}8)$$

then A and B are said to be *statistically independent.* Note that Eq. (1.8-8) and Eq. (1.8-1) imply that

$$P(B/A) = P(B) . \qquad (1.8\text{-}9)$$

which could have been taken equally well as the definition of independence. This concept can be extended as before to cases where more than two variables are involved. Note that Eq. (1.8-8) or (1.8-9) implies that Eq. (1.8-1) can be rewritten for the independent case as

$$P(AB) = P(A)P(B) \qquad (1.8\text{-}10)$$

which also could have been taken as the definition of independence.

Example 1.8

Two dice are thrown. The events A and B are taken to be

Event A - "odd face on first die"

Event B - "odd face on second die"

The probabilities for fair dice are

$$P(A) = 1/2 \qquad P(A/B) = 1/2$$
$$P(B) = 1/2 \qquad P(B/A) = 1/2$$

It is clear that A and B are statistically independent from Eq. (1.8-8) or (1.8-9).

Similarly, three events A, B, and C are said to be *mutually independent* if

$$P(AB) = P(A)P(B)$$
$$P(AC) = P(A)P(C) \qquad (1.8\text{-}11)$$
$$P(BC) = P(B)P(C)$$

and if

$$P(ABC) = P(A)P(B)P(C) \qquad (1.8\text{-}12)$$

The last condition is necessary since the first three do not insure that such events as AB and C are independent as they would be if

$$P\{(AB)(C)\} = P(AB)P(C) = P(A)P(B)P(C) \quad (1.8\text{-}13)$$

Example 1.9

As in Example 1.8, two dice are thrown. The events A, B, and C are taken to be

Event A - "odd face on first die"

Event B - "odd face on second die"

Event C - "sum of the faces odd"

The probabilities for fair dice are

$$P(A) = 1/2 \quad P(A/B) = 1/2 \quad P(A/C) = 1/2$$
$$P(B) = 1/2 \quad P(B/A) = 1/2 \quad P(B/C) = 1/2$$
$$P(C) = 1/2 \quad P(C/A) = 1/2 \quad P(C/B) = 1/2$$

However, A, B, and C cannot occur simultaneously and it is apparent that the events AB and C are not independent.

For N events A_i to be considered statistically independent, it is necessary that, for all combinations $1 \leq i \leq j \leq k \leq \cdots \leq N$,

$$P(A_i A_j) = P(A_i)P(A_j)$$
$$P(A_i A_j A_k) = P(A_i)P(A_j)P(A_k)$$

.

.   (1.8-14)

.

$$P(A_1 A_2 \cdots A_N) = P(A_1)P(A_2)...P(A_N)$$

Consider an event A which must occur in conjunction with one of the mutually exclusive events $B_i, i = 1,2,...N$. In other words,

$$A \subset \bigcup_{i=1}^{N} B_i \qquad (1.8-15)$$

and

$$B_i B_j = \phi \ , \ i,j = 1,2,...,N \ , \ i \neq j \qquad (1.8-16)$$

There may exist situations where the conditional probabilities $P(A/B_i)$ and the probabilities $P(B_i)$ are easily determined directly but the conditional probability $P(B_j/A)$ is desired. It follows from Eq. (1.8-1) that

$$P(B_j/A) = \frac{P(B_j)P(A/B_j)}{P(A)} \ , \ j = 1,2,...,N \qquad (1.8-17)$$

As a consequence of Eqs. (1.8-15) and (1.8-16), the event A may be written as the union of a set of *disjoint* events:

$$A = AB_1 \cup AB_2 \cup \cdots \cup AB_N = \bigcup_{i=1}^{N} AB_i \qquad (1.8-18)$$

Since the probability of the union of a set of disjoint events is the sum of the probabilities of the individual events, this last relationship leads to

$$P(A) = \sum_{i=1}^{N} P(AB_i) = \sum_{i=1}^{N} P(B_i)P(A/B_i) \qquad (1.8-19)$$

This equation may be substituted into Eq. (1.8-17) to yield *Bayes' Formula:*

$$P(B_j/A) = \frac{P(B_j)P(A/B_j)}{\sum\limits_{i=1}^{N} P(B_i)P(A/B_i)} \ , \ j = 1,2,...,N \qquad (1.8-20)$$

It should be emphasized that this formula holds subject to the conditions of Eqs. (1.8-15) and (1.8-16). Much of the abuse of Bayes' formula relates to the lack of validity of these two equations in a given application. Some insight into Eq. (1.8-20) might be gained by examining Fig. 1.4.

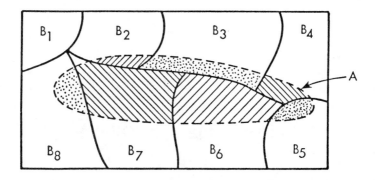

Fig. 1.4 - An Illustration of Bayes' Formula

Example 1.10

A binary (digital) communication channel transmits a signal which is a sequence of 0's and 1's. Each symbol in the sequence is perturbed independently by noise so that, in some cases, errors occur, that is, a 1 is received as a 0 or a 0 as a 1. Let the sent symbols 1 and 0 be designated by B_1 and B_2, respectively, and the received symbols 1 and 0 by A_1 and A_2. The error probabilities are

$P(A_2/B_1) = q_{10} = 1-p_{10} =$ probability 0 was received when 1 was sent.

$P(A_1/B_2) = q_{01} = 1-p_{01} =$ probability 1 was received when 0 was sent.

Let the *a priori* probability that a 1 or a 0 was sent be given respectively by

$$P(B_1) = p \quad , \quad P(B_2) = q = 1-p$$

It is desired to find the *a posteriori* probabilities $P(B_j/A_k)$ where $j, k = 1,2$.

Bayes formula may be applied directly in the form

$$P(B_j/A_k) = \frac{P(B_j)P(A_k/B_j)}{P(B_1)P(A_k/B_1)+P(B_2)P(A_k/B_2)} \quad , \quad j,k = 1,2$$

The four *a posteriori* probabilities are easily written as

$$P(B_1/A_1) = \frac{p\ p_{10}}{p\ p_{10} + (1-p)(1-p_{01})}$$

$$P(B_1/A_2) = \frac{p(1-p_{10})}{p(1-p_{10}) + (1-p)p_{01}}$$

$$p(B_2/A_1) = \frac{(1-p)(1-p_{01})}{p\ p_{10} + (1-p)(1-p_{01})}$$

$$p(B_2/A_2) = \frac{(1-p)p_{01}}{p(1-p_{10}) + (1-p)p_{01}}$$

Note that $P(A_1)$ is just

$$P(A_1) = P(B_1)P(A_1/B_1) + P(B_2)P(A_1/B_2)$$
$$= p\ p_{10} + (1-p)(1-p_{01})$$

and $P(A_2)$ is

$$P(A_2) = P(B_1)P(A_2/B_1) + P(B_2)P(A_2/B_2)$$
$$= p(1-p_{10}) + (1-p)p_{01} = 1 - P(A_1)$$

Consider the special case of a *symmetric* channel where the error probabilities are equal; that is, $q_{10} = q_{01}$ and $p_{10} = p_{01}$. In addition, let the *a priori* probabilities be equal so that $p = q = 1/2$. The result is

$$P(A_1) = P(A_2) = 1/2$$
$$P(B_1/A_2) = P(B_2/A_1) = 1 - p_{10}$$
$$P(B_1/A_1) = P(B_2/A_2) = p_{10}$$

as might be expected.

Recall that two events A and B are statistically independent if and only if Eq. (1.8-3) is satisfied. In this case, the event A and B^c are also independent.

Proof: The probability $P(AB^c)$ can be written as

$$P(AB^c) = P(A) - P(AB)$$

since $A^c = AB \cup AB^c$ and $AB \cap AB^c = \phi$. Thus we have

$$P(AB^c) = P(A) - P(A)P(B) = P(A)[1 - P(B)]$$

or

$$P(AB^c) = P(A)P(B^c) \qquad \bullet \qquad\qquad (1.8\text{-}21)$$

Note that the probability of the event $A \cup B$ can be written as

$$P(A \cup B) = P(A) + P(B) - P(AB) \qquad (1.8\text{-}22)$$

since

$$A \cup B = A + B - AB \qquad (1.8\text{-}23)$$

If A and B are mutually exclusive, then $AB = \phi$ and

$$P(A \cup B) = P(A) + P(B)$$

as previously discussed.

The dependency relationships can be extended to the case where more than two events can occur. Consider the events A, B, and C. If B is replaced in Eq. (1.8-22) by $B \cup C$, the result is

$$P(A \cup B \cup C) =$$
$$P(A) + P(B) + P(C) -$$

$$(1.8\text{-}24)$$

$$P(AB) - P(AC) - P(BC) + P(ABC)$$

where use has been made of the identity

$$A(B \cup C) = AB + AC - ABC. \qquad (1.8\text{-}25)$$

Here $A \cup B \cup C$ means that at least one of the events A, B, or C has occurred and ABC means that all of the events A, B, and C have occurred. In a similar way, Eq. (1.8-24) may be extended to the union of any number of events.

Let B in Eq. (1.8-1) be replaced by BC. Then we have

$$P(ABC) = P(A) \; P(B/A) \; P(C/AB) \qquad (1.8\text{-}26)$$

Equation (1.8-26) may be extended in an obvious manner to yield the joint probability of occurrence of the n events $A_1, A_2, ..., A_n$. The result is the *multiplicative theorem* given by

$$P(A_1 A_2 ... A_n) =$$

$$(1.8\text{-}27)$$

$$P(A_1)P(A_2/A_1)P(A_3/A_1 A_2)...P(A_n/A_1...A_{n-1})$$

In the same way, Eq. (1.8-22) may be extended to yield the probability of occurrence of *at least one* of the events $A_1, A_2, ..., A_n$. The result is the *addition theorem*

$$P\left(\bigcup_{i=1}^{n} A_i\right) =$$

$$(1.8\text{-}28)$$

$$P(A_1) + P(A_1^c A_2) + \cdots + P(A_1^c A_2^c ... A_{n-1}^c A_n)$$

since the sets $A_1, A_1^c A_2, ..., A_1^c A_2^c \cdots A_{n-1}^c A_n$

are disjoint and since their union is equal to $\bigcup\limits_{i=1}^{n} A_i$. Set $n=2$ and note that

$$A_2 = A_1 A_2 + A_1^c A_2 \tag{1.8-29}$$

where $A_1 A_2$ and $A_1^c A_2$ are disjoint. Now Eq. (1.8-28) becomes

$$P(A_1 \cup A_2) = P(A_1) + P(A_2) - P(A_1 A_2) \tag{1.8-30}$$

in agreement with Eq. (1.8-22).

The *covering theorem* is easily obtained from Eq. (1.8-28). Note that

$$P(A_1^c A_2) \leq P(A_2) \tag{1.8-31}$$

$$P(A_1^c A_2^c A_3) \leq P(A_3) \tag{1.8-32}$$

$$\cdot\cdot$$
$$\cdot\cdot$$
$$\cdot\cdot$$

$$P(A_1^c A_2^c ... A_{n-1}^c A_n) \leq P(A_n) \tag{1.8-33}$$

since $A_1^c A_2^c ... A_{n-1}^c A_n \subseteq A_n$ for all $n \geq 1$. It follows now from Eq. (1.8-28) and these last three expressions that

$$P(\bigcup_{i=1}^{n} A_i) \leq \sum_{i=1}^{n} P(A_i) \tag{1.8-34}$$

for any denumerable class of sets $\{A_i \mid i=1,2,...,n\}$ whether or not the A_i are disjoint.

Suppose that a definition or relation is valid in the sample space S except on an *empty* or *null* set (i.e. on a set of probability zero); then the definition or relation is said to be valid *almost everywhere* (a.e.), or *almost surely* (a.s.), or with *probability one* (prob. 1).

PROBLEMS

1. Indicate by cross-hatching the following sets on the Venn diagram.

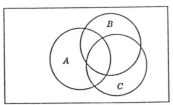

(a) $B \cap C$ (b) $A \cap B^c$
(c) $(A \cup B^c) \cap C^c$ (d) $(A^c \cap B)^c \cup C$

2. Let A be the set of positive even integers, let B be the set of positive integers divisible by 3, and let C be the set of positive odd integers. Describe the following sets.

<ul style="list-style:none">
<li>(a) $A \cup C$</li>
<li>(b) $A \cap B$</li>
<li>(c) $A \cap C$</li>
<li>(d) $(A \cup B) \cap C$</li>
<li>(d) $A \cup (B \cap C)$</li>
</ul>

3. Using the basic definitions of set equality, intersection, and union, prove the distribution law

$$A \cap (B \cup C) = (A \cap B) \cup (A \cap C)$$

4. Prove that, if $B \supset A$, then $B^c \subset A^c$. Make a Venn diagram to illustrate your proof.

5.

$$\text{If } \binom{n}{13} = \binom{n}{6} \text{ , what is n?}$$

$$\text{If } \binom{13}{r} = \binom{13}{r-3} \text{ , what is r?}$$

6. Three identical coins, marked **A**, **B**, and **C** for identification, are tossed simultaneously. If the order in which the coins fall is noted, how many ways can the coins fall assuming each coin has a "head" and "tail" side?

7. How many distinguishable arrangements are there of ten persons (a) in a row, (b) in a ring?

8. The quantity $B - A$ is sometimes called the *relative complement* of A with respect to B. (a) Illustrate with Venn diagrams the two situations $B \supset A$ and $B \not\supset A$. (b) Show that $P\{B - A\} = P\{B\} - P\{AB\}$.

9. Three dice are rolled. Let A be the event that at least one ace appears and let B be the event that no two dice show the same value. Find $P(A)$, $P(B)$, and $P(AB)$. Are A and B statistically independent? Prove your answer.

10. Show that the set function $P\{\cdot\}$ is *monotone*; that is, if $A \subset B$, then $P\{A\} \le P\{B\}$.

11. Consider a game which consists of two successive trials. The first trial has outcomes A or B and the second outcomes C or D. The probabilities for the four possible outcomes of the game are as follows:

Outcome	AC	AD	BC	BD
Probability	$\dfrac{1}{3}$	$\dfrac{1}{6}$	$\dfrac{1}{6}$	$\dfrac{1}{3}$

Are A and C statistically independent? Prove your answer.

12. Consider a family with two children of different ages. Assume each child is as likely to be a boy as it is to be a girl. What is the conditional probability that both children are boys, given that (a) the older child is a boy, and (b) at least one of the children is a boy?

13. The membership of a certain club is composed of 30 men and 20 women. If 40% of the men and 60% of the women play bridge, what is the conditional probability that a member who plays bridge is a man?

14. Show that $P\{B/A\} \geq P\{AB\}$ and that $P\{B/A\} = 0$ if $AB = \phi$.

15. The network of switches A, B, C, D are connected across the power lines x, y as shown. Each switch has probability p of not closing when operated, and each switch functions independently of the other switches. What is the probability that the circuit from x to y will fail to close when all four switches are operated? Will the addition of bus m change this probability? If so, what is the new probability?

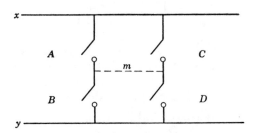

16. The three letters in the word "cat" are permuted in all possible distinct ways and then one permutation is picked at random. What is the probability that it spells a recognizable word in the English language?

17. What is the probability of throwing a "4" before a "7" with two dice?

18. A random experiment has k possible outcomes $E_1, E_2, ..., E_k$ which are mutually exclusive and exhaustive with probabilities $p_1, p_2, ..., p_k$. The experiment is repeated n times. Find the probability $p(r_1, r_2, ..., r_k)$ of obtaining exactly r_1 occurrences of event E_1, r_2 of event E_2, ..., and r_k of E_k where $r_1 + r_2 + \cdots + r_k = n$. This probability $p(r_1, r_2, ..., r_k)$ is called the *multinomial distribution*. When $k = 2$, the result is the *binomial distribution*.

19. In a common gambling game, three people each toss a coin and compare the results. The "odd" man wins all three coins; if all the coins are alike, the toss is repeated. Show that the game is "fair"; that is, that each man has an equal

chance of winning. In a sequence of ten games, find the probability that a given one of the players will win *exactly* eight times. Ignore those games where all the coins are alike. Find the probability that this player will win *at least* eight times in the ten games.

20. When a fair die is rolled, find the probability that the number of spots on the upper face is greater than r where $r = 1, 2, 3, 4, 5$.

21. Success in playing poker rests partly on a correct knowledge of the probabilities of drawing various winning hands. For a five-card hand drawn from an ordinary bridge deck, find the probability of (a) one pair, (b) two pairs, (c) three of a kind, (d) five cards of the same suit (flush), (e) three cards of one kind and two of another (full house), (f) five cards in numerical sequence (a straight), (g) four of a kind, and (h) a straight flush.

22. You are one of r people where $r > 1$. A suicidal mission is to be assigned to the person who draws the short straw from a set of r straws. What place in line should you take to maximize your chances of survival?

23. A group of students consists of 60% men and 40% women. Among the men, 25% are blond while among the women, 45% are blond. A person is chosen at random from the group and is found to be blond. Use Bayes formula to compute the probability that the person is a man.

Chapter 2

RANDOM VARIABLES

2.1 *Definition* - In Chapter 1 we developed the concept of a probability space which completely describes the outcome of a random experiment. Fundamental to the specification is a set of elementary outcomes or points $\omega \in S$ where S is the sample space associated with the probability space $(S, \mathbf{B}, P)$. These elementary outcomes are the building blocks from which the events are constructed. As was emphasized earlier, the elementary outcomes can be any objects whatsoever. Certainly, they need not be numbers. In the tossing of a coin, for example, the outcomes will usually be "heads" or "tails". On the other hand, if a single die is tossed, one possible set of outcomes are the numbers 1,2,3,4,5, and 6 corresponding to the number of dots on the upper face of the die.

It is usually convenient, however, to associate a real number with each of the elementary outcomes (or elements) ω of the sample space S. The result is a mapping of the space S to the real line R in the manner indicated in Fig. 2.1. The assignment of the real numbers amounts to defining a real-valued function on the elements of the sample space; this function is called a *random variable*. It is desired that this random variable assume each of its possible values with a definite probability; therefore we are led to the following equivalent definitions of a random variable:

Definition 1(a): A *random variable* X is a real-and single-valued function defined on the points (elementary outcomes) of a probability space. To each point, there corresponds a real number, the *value* of the random variable at that point. For every real number x, the set $\{X = x\}$, the set of points at which X assumes the value x, is an event. Also, for every pair of real numbers x_1 and x_2, the sets $\{x_1 < X < x_2\}$, $\{x_1 \leq X \leq x_2\}$, $\{x_1 < X \leq x_2\}$, $\{x_1 \leq X < x_2\}$, $\{X < x_1\}$, $\{X \leq x_1\}$, $\{X < x_2\}$, and $\{X \leq x_2\}$ are events.

Definition 1(b): A *random variable* X on a probability space $(S, \mathbf{B}, P)$ is a real-and single-valued function whose domain is S and which is $\mathbf{B}$ -*measurable;* that is, for every real number x,

$$\{\omega \in S \mid X(\omega) \leq x\} \in \mathbf{B}.$$

Several comments should be made about these definitions. In Definition 1(a), if S is a finite space, then the condition of the last sentence is automatically satisfied; that is, every set of points is an event.

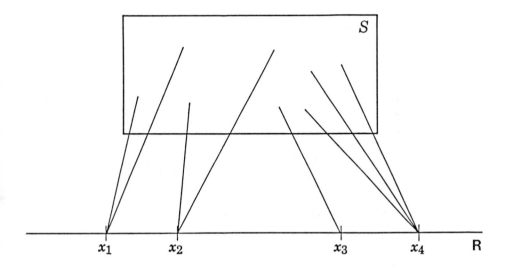

Fig. 2.1 - Random variables as mappings

In terms of Definition 1(b), all subsets of a finite probability space are **B** -*measurable*. The notation used in Definition 1(b) is somewhat cumbersome and it will be convenient to express the set in curly brackets by $\{X \leq x\}$, that is, we shall use the following notation:

$$\{X \leq x\} = \{\omega \in S \mid X(\omega) \leq x\} \qquad (2.1\text{-}1)$$

and, for any subset **B** of **R**,

$$\{X \in \mathbf{B}\} = \{\omega \in S \mid X(\omega) \in \mathbf{B}\} \qquad (2.1\text{-}2)$$

One of the simplest examples of a random variable is the *indicator* or *indicator function* I_A of an event A. If $A \in \mathbf{B}$ (that is, if A is an event), then I_A is a function taking on the value unity at all points of A and the value zero at all other points; thus

$$I_A(\omega) = \begin{cases} 1 \,, \omega \in A \\ 0 \,, \omega \notin A \end{cases} \qquad (2.1\text{-}3)$$

Every event has an indicator and every function assuming only the values unity and zero is the indicator of some event. The indicator for the event A is shown schematically in Fig. 2.2. For an arbitrary function f, the product $I_A f$ is another function taking on the values of f on A and zero elsewhere.

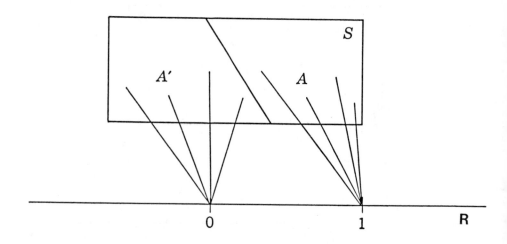

Fig. 2.2 - The mapping from the basic space S to the real line R produced by the indicator function $I_A (\bullet)$.

It is clear that the concept of a random variable has been introduced in order to map each element of the abstract sample space S onto the real line R. The domain of X is S and the range is R. Thus subsets of the sample space become sets of points on the real line. To every set A in S there corresponds a set in R, called the *image* of A and denoted by $X(A)$. Also, for every set T in R, there exists in S the *inverse image* $X^{-1}(T)$ where

$$X^{-1}(T) = \{\omega \in S \mid X(\omega) \in T\} \qquad (2.1\text{-}4)$$

The function X^{-1} is called the *inverse function* of X and maps the real line R into the abstract space S. Since X was defined to be single-valued, the inverse function X^{-1} maps disjoint sets T_1 and T_2 in R into disjoint sets in S as illustrated in Fig. 2.1 and Fig. 2.2. Thus, the inverse functions preserve all set operations, and, if a class of sets $\mathbf{F}$ in R is closed under a given set operation, the class $X^{-1}(\mathbf{F})$ in S is closed under the same operation, where the class $X^{-1}(\mathbf{F})$ is defined by

$$X^{-1}(\mathbf{F}) = \{X^{-1}(T) \subset S \mid T \in \mathbf{F}\} \qquad (2.1\text{-}5)$$

Given a probability space $(S,\mathbf{B},P)$, it is clear that the random variable X defined on this space induces on its range space R a new probability space $(R,\mathbf{F},P_X)$ where the new probability assignment P_X is defined by

$$P_X(T) = P(X^{-1}(T)) \qquad (2.1\text{-}6)$$

It is also clear that, to every event $A \in S$ with associated probability $P(A)$, there corresponds an event $T \subset R$ with equal probability $P_X(T)$, that is,

$$P_X(T) = P(A) \quad , \quad X(A) = T \text{ or } X^{-1}(T) = A \quad (2.1\text{-}7)$$

Thus it is no longer necessary to consider the basic space $(S, \mathbf{B}, P)$; the new space $(R,\mathbf{F},P_X)$ is equivalent and its sample points x are real numbers.

Let us now consider several rather simple examples of random variables.

Example 2.1

A random experiment consists of tossing a coin three times. Assume that the coin is weighted so that the probability of a head on each toss is $2/3$ and of a tail is $1/3$. The random variable X will be taken to be the number of heads produced in the three tosses. Thus X takes on the four values $x_0 = 0$, $x_1 = 1$, $x_2 = 2$, and $x_3 = 3$. The sample space and probabilities $P(x_i)$ may be tabulated as:

ω	$X(\omega)$	$P(\{\omega\})$	
HHH	3	8/27	$P(X=3) = 8/27$
HHT	2	4/27	
HTH	2	4/27	$P(X=2) = 12/27$
THH	2	4/27	
HTT	1	2/27	
THT	1	2/27	$P(X=1) = 6/27$
TTH	1	2/27	
TTT	0	1/27	$P(X=0) = 1/27$

Example 2.2

A marksman is given four cartridges and is told to fire at a target until he has hit it or has used up the four cartridges. The probability of a hit on each shot is p and of a miss is $q = 1-p$, independently of other shots. Let the number of used cartridges be the random variable X taking on values 1,2,3, and 4. Find the function* $P\{X=x\}$, $x=1,2,3,4$; this is the *(probability) density function* of the random variable X.

The probabilities are easily tabulated; we have:

$P\{X=1\} = p$:by definition.

$P\{X=2\} = qp = (1-p)p$:miss with first shot; hit with second.

$P\{X=3\} = q^2p = (1-p)^2p$:miss with first two shots; hit with third.

$P\{X=4\} = q^3 = (1-p)^3$:miss with first three shots; doesn't matter whether fourth shot is hit or miss.

Note the sum of these probabilities must be unity since the associated events exhaust the sample space; that is,

$$P(S) = P\{X=1\}+P\{X=2\}+P\{X=3\}+P\{X=4\}$$
$$= p + (1-p)p + (1-p)^2p + (1-p)^3 = 1$$

Suppose now that the supply of cartridges is unlimited and the marksman fires at the target until he hits? In this case the probabilities are

$$P\{X=n\} = (1-p)^{n-1}p \quad , \quad n=1,2,...$$

Again the sum must be unity:

$$\sum_{n=1}^{\infty} P\{X=n\} = \sum_{n=1}^{\infty} (1-p)^{n-1}p = \frac{p}{1-(1-p)} = 1$$

Example 2.3

A fair coin is tossed; two possible outcomes, "heads" and "tails", exist each with probability one-half. A random variable X is defined by assigning the value unity to the occurrence of a head and the value zero to the occurrence of a tail; thus

$$P\{X=x\} = \begin{cases} 1/2, & x=0 \\ 1/2, & x=1 \\ 0, & \text{elsewhere} \end{cases}$$

---

*To simplify notation, we will suppress the parentheses in expressions such as $P(\{X=x\})$.

Another well-defined event is the event $X \leq x$ for $-\infty < x < \infty$ with probability

$$P\{X \leq x\} = \begin{cases} 0 \,, & x < 0 \\ 1/2, & 0 \leq x < 1 \\ 1 \,, & x \geq 1 \end{cases}$$

The function $P\{X=x\}$ and $P\{X \leq x\}$ are plotted in Fig. 2.3. The first is called the *(probability) density (or mass) function* and the second is called the *(cumulative) distribution function* of the random variable X.

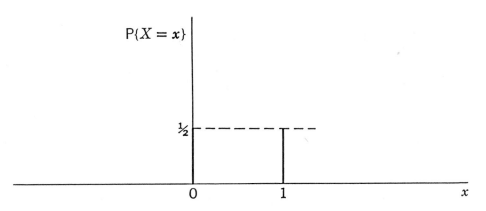

(a) The density function

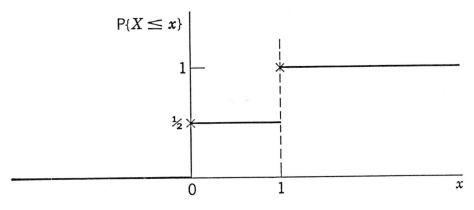

(b) The (cumulative) distribution function

Fig. 2.3 - A coin-tossing experiment.

2.2 *Discrete Random Variables* - In the examples just considered, the random variable has taken on only a discrete set of values. In such cases we say that we are dealing with a *discrete random variable,* sometimes denoted as X_i where the subscript i indexes the possible values of X_i. More precisely, *a random variable X will be said to be discrete* if there is a denumerable sequence of distinct numbers x_i such that

$$P\left\{\bigcup_i \{X=x_i\}\right\} = \sum_i P\{X=x_i\} = 1 \qquad (2.2\text{-}1)$$

If X is discrete, it can be written in terms of indicators as

$$X = \sum_i x_i \, I_{\{X=x_i\}} \qquad (2.2\text{-}2)$$

Example 2.4

Let A, B, and C be three mutually exclusive and exhaustive events. Let a random variable X be defined which associates the number 1 with event A, the number 2 with event B, and the number 3 with with event C. This random variable may be written as

$$X(\omega) = 1 \times I_A(\omega) + 2 \times I_B(\omega) + 3 \times I_C(\omega)$$

where $\omega \in S$. For example, when $\omega \in B$, $X(\omega) = 2$.

In accordance with the definition, a discrete random variable takes on only a denumerable number of possible values x_i, $i=1,2,\dots$. As mentioned in Examples 2.2 and 2.3, the *density function* $f_X(x_i)$ of a discrete random variable X is determined by this set of numbers x_i and their associated probabilities; that is,

$$f_X(x_i) = P\{X=x_i\} \ , \quad i=1,2,\dots \qquad (2.2\text{-}3)$$

It is clear that

$$f_X(x_i) \geq 0 \qquad (2.2\text{-}4)$$

and that

$$\sum_i f_X(x_i) = 1 \qquad (2.2\text{-}5)$$

As in Example 2.3, it will sometimes be convenient to treat the function f as though it were defined for all real x. In this case we will write

$$f_X(x) = P\{X=x\} \qquad (2.2\text{-}6)$$

and consider that $f_X(x)$ is zero whenever x is not one of the set of numbers $\{x_i\}$.

The probabilities of such events as $\{a < X < b\}$ are easily calculated since they are given by the sum

$$P\{a < X < b\} = \sum_{a < x_i < b} f_X(x_i) \qquad (2.2\text{-}7)$$

In particular, the probability of the event $\{X \leq x\}$ will be denoted by $F_X(x)$ and is given by

$$F_X(x) = P\{X \leq x\} = \sum_{x_i \leq x} f_X(x_i) \qquad (2.2\text{-}8)$$

As previously mentioned, the function $F_X(x)$ is called the *(cumulative) distribution function* (c.d.f.) of the random variable X. Various properties of this function will be developed later.

A discrete random variable whose range consists of a finite set of values is sometimes called a *simple random variable*. If the range consists of a set which is denumerably infinite, the random variable is sometimes called *elementary*. Thus a discrete random variable is either simple or elementary. Note that an indicator function is an example of a simple random variable.

2.3 *Continuous Random Variables* - Not all random variables are discrete. For example, X may assume a continuum of values in some interval as the following example indicates.

Example 2.5

Consider the *uniform distribution* where the range of the random variable is some finite interval (a, b). Let (x_1, x_2) be any interval inside (a, b). The probability that the continuous random variable X lies in this interval is assumed to be proportional to the length of the interval, that is,

$$P\{x_1 < X < x_2\} = \lambda(x_2 - x_1) \ , \ \ a \leq x_1 < x_2 \leq b$$

The probability that X lies outside this interval is zero. By Axiom 1 of Chapter 2, $P(A) \geq 0$ for any event A; consequently $\lambda \geq 0$. By Axiom 2, we have

$$P(S) = 1 = P\{a < X < b\} = \lambda(b - a) = 1$$

or

$$\lambda = \frac{1}{b - a}$$

Therefore the probability that X lies in the interval (x_1, x_2) is

$$P\{x_1 < X < x_2\} = \frac{x_2 - x_1}{b - a} \ , \ \ a \leq x_1 < x_2 \leq b$$

Note that this last expression can be rearranged as

$$\frac{P\{x_1 < X < x_2\}}{x_2 - x_1} = \frac{1}{b-a} \ , \quad a \leq x_1 < x_2 \leq b$$

In this form it is called the *(probability) density function* for the uniformly distributed random variable X and is shown in Fig. 2.4(a).

Let us calculate now the probability $P\{X \leq x\}$ where x is any real number. If $x < a$, it is clear that this probability is zero; in the same way, if $x > a$, the probability must be unity. For $a \leq x \leq b$, we have

$$P\{X \leq x\} = P\{a < X \leq x\} = \frac{x-a}{b-a} \ , \quad a \leq x \leq b$$

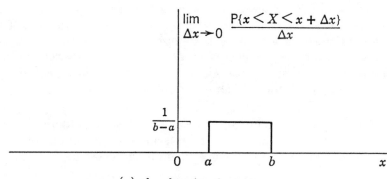

(a) the density function

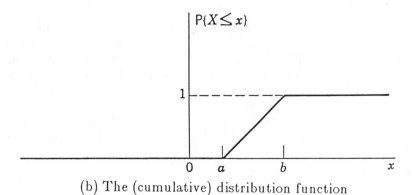

(b) The (cumulative) distribution function

Fig. 2.4 - The uniform distribution.

The probability $P\{X \leq x\}$ is called the *(cumulative) distribution function* for the uniformly distributed random variable X and is plotted in Fig. 2.4(b).

In general, *a random variable X will be said to be continuous* if $P\{X=x\} = 0$ for all x and if a function $f_X(x)$ exists such that, for all x,

$$P\{x<X<x+\Delta x\} = \int_{x}^{x+\Delta x} f_X(y)\ dy \qquad (2.3\text{-}1)$$

The function $f_X(x)$ is called the *density function* of the continuous random variable X. From an alternative point of view, if $f_X(x)$ is continuous at x, it could be defined as

$$f_X(x) = \lim_{\Delta x \to 0} \frac{P\{x<X<x+\Delta x\}}{\Delta x} \qquad (2.3\text{-}2)$$

It follows from Eq. (2.3-1) and the Mean-Value Theorem that, if $f_X(x)$ is continuous at x,

$$P\{x<X<x+dx\} \approx f_X(x)\ dx \qquad (2.3\text{-}3)$$

The quantity $f_X(x)dx$ will be called the *probability differential* $dF_X(x)$. Now the probability of the event $\{x_1<X<x_2\}$ may be written as

$$P\{x_1<X<x_2\} = \int_{x_1}^{x_2} f_X(x)\ dx = \int_{x_1}^{x_2} dF_X(x) \qquad (2.3\text{-}4)$$

If $f_X(x)$ exists; that is, if $f_X(x) < \infty$, Eq. (2.3-1) implies that the probability of the event $\{X=x\}$ is zero since

$$P\{X=x\} = \lim_{\Delta x \to 0} P\{x<X<x+\Delta x\} =$$

$$(2.3\text{-}5)$$

$$\lim_{\Delta x \to 0} f_X(x)\Delta x = 0$$

It should be kept in mind that this last expression does not preclude the occurrence of the event $\{X=x\}$. It was pointed out previously that the probability of the null event is zero but that the converse is not true.

The probability of the event $\{X \leq x\}$ may be written as

$$P\{X \leq x\} = \int_{-\infty}^{x} f_X(y)\ dy = \int_{-\infty}^{x} dF_X(y) \qquad (2.3\text{-}6)$$

or, from the fundamental theorem of the calculus,

$$f_X(x) = \frac{d}{dx} F_X(x) \qquad (2.3\text{-}7)$$

where $F_X(x)$ is the distribution function of X.

It is clear that, for a continuous random variable X, a continuum of values is permissible over some interval or set of intervals.

In this case, the probability of a single selected value of X occurring is zero, as shown by Eq. (2.3-5). The situation is analogous to considering the area under a simple curve. The area under any point is zero but the area under any segment of nonzero length may be nonzero. We are led to consider not the probability that a random variable X takes on some value x but rather that it lies in some interval $a \leq X \leq b$; that is, $P\{a \leq X \leq b\}$.

Thus, a one-to-one correspondence does not exist between the point values of a continuous random variable and a denumerable set of underlying events $A, B, C, \ldots$ even if an infinite number of events are considered since the points in a continuum are not denumerable. The problem has already been resolved, however, in an elementary fashion. Divide the real line into two parts by the number x and consider values of the random variable $X \leq x$ and $X > x$. With an arbitrary event A associate the values of the random variable $X \leq x$. Thus the probability $P(A)$ of the event A becomes the probability $P\{X \leq x\}$. This probability has already been denoted by $F_X(x)$ and is the distribution function of X. If this function is given for all values of x, $-\infty < x < \infty$, then the values of the random variable have been defined.

2.4 *Random Vectors* - The concept of a random variable may be extended in an obvious way to more than one dimension.

Definition 2: A *random vector* $\mathbf{X} = (X_1, X_2, \ldots, X_n)$ is a vector of real-valued functions on S such that the probability of the event $\{X_1 \leq x_1, X_2 \leq x_2, \ldots, X_n \leq x_n\}$ is defined for all n-tuples $x_1, x_2, \ldots, x_n$.

A random vector will also be called an *n-dimensional random variable* or a *vector random variable*. The same type of notation will be used as in one dimension, that is

$$\{\mathbf{X} \in R\} = \{\omega \in S \mid [X_1(\omega), \ldots, X_n(\omega)] \in R\}$$

Example 2.6

In the tossing of a fair coin, assign the value unity to the occurrence of a head and the value zero to the occurrence of a tail. Toss three coins and define the random vector $\mathbf{X} = (X_1, X_2, X_3)$ where X_i is the outcome of the toss of the i-th coin. The random vector $\mathbf{X}$ can assume the eight values $(0,0,0)$, $(0,0,1)$, $(0,1,0)$, $(1,0,0)$, $(1,1,0)$, $(1,0,1)$, $(0,1,1)$, and $(1,1,1)$. A *trivariate distribution function* $F_{\mathbf{X}}(x_1, x_2, x_3)$ can be defined by

$$F_{\mathbf{X}}(x_1, x_2, x_3) = P\left\{\bigcap_{i=1}^{3} \{X_1 \leq x_i\}\right\} = F_{\mathbf{X}}(\mathbf{x})$$

where $-\infty < x_i < \infty$ for $1 \leq i \leq 3$. This distribution function is easily found to be

$$F_{\mathbf{X}}(\mathbf{x}) = \begin{cases} 0 & , \ x_1 < 0 \text{ or } \ x_2 < 0 \text{ or } x_3 < 0 \\ 1/8, & 0 \leq x_1 < 1, \ 0 \leq x_2 < 1, \ 0 \leq x_3 < 1 \\ 1/4, & 0 \leq x_1 < 1, \ 0 \leq x_2 < 1, \ x_3 \geq 1 \text{ or} \\ & \quad 0 \leq x_1 < 1, \ x_2 \geq 1, \ 0 \leq x_3 < 1 \text{ or} \\ & \quad x_1 \geq 1, \ 0 \leq x_2 < 1, \ 0 \leq x_3 < 1 \\ & \quad \cdot \quad \cdot \quad \cdot \\ & \quad \cdot \quad \cdot \quad \cdot \\ & \quad \cdot \quad \cdot \quad \cdot \\ 1 & , \ x_1 \geq 1, \ x_2 \geq 1, \ x_3 \geq 1 \end{cases}$$

The interested reader can fill in the missing values.

It follows straightforwardly from the definition of a random vector that if $X_1, X_2, ..., X_n$ are n random variables, then $\mathbf{X} = (X_1, X_2, ..., X_n)$ is a random vector.

2.5 *Independence of Random Variables* - Recall from Eq. (1.8-3) that two events A and B are *independent* if and only if

$$P(AB) = P(A) \ P(B) \qquad (2.5\text{-}1)$$

Let $\mathbf{A} = \{A_\lambda, \lambda \in \Lambda\}$ be a set of events where Λ is the indexing set. As discussed in Section 1.8, Eq. (1.8-7), these events are independent if and only if

$$P(A_{\lambda_1} \cap \cdots \cap A_{\lambda_n}) = \mathop{\pi}_{i=1}^{n} P(A_{\lambda_i}) \qquad (2.5\text{-}2)$$

for every positive integer n and every n distinct elements $\lambda_1, ..., \lambda_n$ in the set Λ.

In the same way, let $\mathbf{X} = \{X_\lambda, \lambda \in \Lambda\}$ be a set of random variables indexed by λ. These random variables are independent provided the underlying events

$$A_{\lambda_i} = \{\omega \in S \mid X_{\lambda_i}(\omega) \in T_i\}$$

are independent [i.e., satisfy Eq. (2.5-2)] for every positive integer n, for every choice of $\mathbf{F}$-measurable sets $T_1, ..., T_n$ in R_n, and for every n distinct elements $\lambda_1, ..., \lambda_n$ in the set Λ.

Consider the case where X_1 and X_2 are two discrete random variables with possible sets of values $\{x_{1k}\}$ and $\{x_{2j}\}$. Then the definition of the last paragraph is equivalent to saying that X_1 and X_2 are independent if and only if the events $\{X_1 = x_{1k}\}$ and $\{X_2 = x_{2j}\}$ are independent for all k and j. Consider again two random variables X_1 and X_2 which are not necessarily discrete. As pointed out in Eq. (2.1-4) the inverse images $X_1^{-1}(T_1)$ and $X_2^{-1}(T_2)$ map sets in R_2 into events in the abstract space S. The random variables X_1 and X_2 are independent if and only if the events $X_1^{-1}(T_1)$ and $X_2^{-1}(T_2)$ are independent events for every T_1 and T_2.

Example 2.7

Consider the indicator functions I_A and I_B. These discrete random variables are independent if and only if the events A and B are independent. Note that $I_A^{-1}(T_1)$ is one of the sets in the sigma-field $\mathbf{B}_1 = \{A, A^c, S, \Phi\}$ and $I_B^{-1}(T_2)$ is one of the set in the sigma-field $\mathbf{B}_2 = \{B, B^c, S, \Phi\}$. If A and B are independent, it is easy to show that all the other members of $\mathbf{B}_1$ are independent of all of the other members of $\mathbf{B}_2$. For example

$$P(AB^c) = P(A) - P(AB) = P(A) - P(A)P(B)$$
$$= P(A)[1-P(B)] = P(A)P(B^c)$$

and A and B^c are independent if A and B are independent. In the same way

$$P(A\Phi) = P(\Phi) = 0 = P(A)P(\Phi)$$

and

$$P(AS) = P(A) = P(A)P(S)$$

Thus I_A and I_B are independent.

PROBLEMS

1. Write out completely the function $F_X(x)$ of Example 2.6.

2. Let A and B be two events such that $A \in \mathbf{B}$ and $B \in \mathbf{B}$ and let I_A and I_B be the indicators of these sets. Show that
   a) $I_A I_B = \min\{I_A, I_B\} = I_{AB}$
   b) $I_{A \cup B} = \max\{I_A, I_B\} = I_A + I_B - I_{AB}$
   c) $I_{A \cup B} = I_A + I_B$ iff. A and B are disjoint

3. Let X by a simple random variable given by

$$X(\omega) = 2 \bullet I_A(\omega) - 3 \bullet I_B(\omega) + 5 \bullet I_C(\omega) + 1 \bullet I_D(\omega)$$

where A, B, C, and D are mutually disjoint events such that $A \cup B \cup C \cup D = S$. Determine the range of X. Find those events belonging to the set $X^{-1}(T)$ when T is the following intervals:

a) $T = (1,2]$ d) $T = (2,5]$

b) $T = (-\infty, 4)$ e) $T = [2,5)$

c) $T = [0, \infty)$ f) $T = (1, \infty)$

4. A man stands at the origin of a one-dimensional coordinate system. He tosses a coin n times (n is finite). After each toss, he moves one unit to the right if the toss yielded a head and one unit to the left if it yielded a tail. Let A_i be the event that a head appeared on the i-th toss. Write an expression in terms of the indicators I_{A_i} and $I_{A_i^c}$ for the man's position at the end of n tosses. Find a general expression for his possible positions after n tosses and the probability of each.

5. A fair die is rolled until a two appears. Let the random variable X be the number of rolls required. Find the c.d.f. $P\{X \leq x\}$. Repeat for the case where two fair dice are rolled.

6. A fair die is rolled until every face has appeared at least once. Let the random variable X be the number of rolls. Find the c.d.f. $P\{X \leq x\}$ and the density function $P\{X = x\}$.

Chapter 3

DISTRIBUTION AND DENSITY FUNCTIONS

3.1 *Distribution Functions* - In Chapter 2, the distribution function (or cumulative distribution function) has been introduced both explicitly and implicitly. We repeat the definition.

Definition:

If X is a random variable, its (cumulative) distribution function F_X is defined by

$$F_X(x) = P\{X \le x\} \quad \text{for all } x \in (-\infty, \infty) \qquad (3.1\text{-}1)$$

Note that F_X is well-defined since the definition of a r.v. requires that $\{X \le x\}$ is an event. It is clear that two different random variables can have the same distribution function.

Example 3.1

Let the sample space S consist of two elementary events so that $S = H, T$ with probabilities

$$P(H) = \frac{1}{2} \quad , \quad P(T) = \frac{1}{2}$$

Define two random variables X and Y by

$$X(H) = 0 \quad , \quad X(T) = 1$$

and

$$Y(H) = 1 \quad , \quad Y(T) = 0$$

The distribution functions F_X and F_Y are given by

$$F_X(x) = F_Y(x) = \begin{cases} 0 & , \ x < 0 \\ 1/2 & , \ 0 \le x < 1 \\ 1 & , \ x \ge 1 \end{cases}$$

and is illustrated in Fig. 2.2(b).

Two random variables are said to be *equivalent* if they have the same distribution function. It will be shown later that to every distribution function there corresponds a random variable which is uniquely determined up to equivalence.

3.2 *Properties of Distribution Functions* - A number of fundamental properties of distribution functions follow directly from the definition of Eq. (3.3-1).

(1) $F_X(x)$ *is monotone nondecreasing* ; that is, $F_X(x+h) \geq F_X(x)$ if $h > 0$.

Proof: The function $F_X(x)$ is a probability by definition. Associate the event A with the probability $P\{X \leq x\}$ and the event B with the probability $P\{X \leq x+h\}$ where $h > 0$. Then $B \supseteq A$ and $P(B) \geq P(A)$, or $F_X(x+h) \geq F_X(x)$. •

(2) $F_X(x)$ *is right-continuous* ; that is,

$$\lim_{\substack{h \to 0 \\ h > 0}} F_X(x+h) = F_X(x) \text{ for all } x.$$

Proof: From the definitions of the events A and B in (1), we have

$$P\{x < X \leq x+h\} = F_X(x+h) - F_X(x) \tag{3.2-1}$$

or

$$\lim_{\substack{h \to 0 \\ h > 0}} P\{x < X \leq x+h\} = P(\Phi) = 0 \tag{3.2-2}$$

or

$$0 = \lim_{\substack{h \to 0 \\ h > 0}} F_X(x+h) - F_X(x) \quad • \tag{3.2-3}$$

(3)

$$\lim_{x \to -\infty} F_X(x) = 0 \quad , \quad \lim_{x \to \infty} F_X(x) = 1 \tag{3.2-4}$$

Proof: We will assume that the random variable X can take on only finite values; then

$$\lim_{x \to -\infty} F_X(x) = P(\Phi) = 0 \tag{3.2-5}$$

and

$$\lim_{x \to \infty} F_X(x) = P(S) = 1 \tag{3.2-6}$$

We shall continue to assume, unless specifically stated otherwise, that the random variable X takes on only finite values and shall use the notation

$$\lim_{x \to -\infty} F_X(x) = F_X(-\infty) \quad , \quad \lim_{x \to \infty} F_X(x) = F_X(\infty) \qquad (3.2\text{-}7)$$

$$\lim_{\substack{h \to 0 \\ h > 0}} F_X(x + h) = F_X(x^+) = F_X(x + 0) \qquad (3.2\text{-}8)$$

$$\lim_{\substack{h \to 0 \\ h > 0}} F_X(x - h) = F_X(x^-) = F_X(x - 0) \qquad (3.2\text{-}9)$$

There are a number of corollaries to these three properties which are either obvious or follow immediately from the previous discussion:

$$0 \leq F_X(x) \leq 1 \qquad (3.2\text{-}10)$$

$$F_X(b) - F_X(a) = P\{a < X \leq b\} \qquad (3.2\text{-}11)$$

$$F_X(x) - F_X(x^-) = P\{X = x\} \qquad (3.2\text{-}12)$$

$$F_X(x^-) = P\{X < x\} \qquad (3.2\text{-}13)$$

$$1 - F_X(x) = P\{X > x\} \qquad (3.2\text{-}14)$$

It is apparent from Corollary (3) that the distribution function is continuous for the continuous random variable X. For a discrete random variable, the distribution function is constant except for jumps at a denumerable set of points $\{x_i\}$, the jumps being equal to the probabilities $P\{X = x_i\}$. These characteristics will be discussed more precisely in the next section. They are illustrated in Figs. 2.2 and 2.3 and in the following example:

Example 3.2

In Example 2.1, a random variable was defined. The distribution function for this random variable is shown in Fig. 3.1. The crosses on the graph are used to indicate the right-continuity of the function $F_X(x)$.

———————

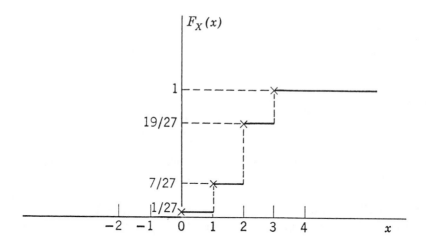

Fig. 3.1 - A cumulative distribution function for
a discrete random variable.

3.3 *The Decomposition of Distribution Functions* - Let F be a
real-valued function defined on the entire real line $(-\infty,\infty)$ and let
it satisfy the three basic properties (1), (2), and (3) of Section 3.2.
We begin by proving that F has at most a denumerable number
of discontinuities and these are jumps. (Note: a jump at a point
in a function is often called a *saltus*).

Proof: Since the function F is right-continuous, it is discontinuous
at a point x_0 iff. $F(x_0^-) < F(x_0)$; the difference $F(x_0) - F(x_0^-)$ will
be called the *jump* $p(x_0)$ at x_0. Let n be a positive integer. The
function $F(x)$ can have at most n points where the jump is $1/n$ or
more; otherwise $F(\infty)-F(-\infty)$ will be greater than unity. Thus the
jumps can be ordered in the following way. There can be no more
than one point with jump which is at least one; there can be no
more than two additional jumps which lie between $1/2$ and $1;...,$
there are no more than n additional jumps which lie between $1/n$
and $1/(n-1)$. The jumps can be numbered $x_1, x_2,...,x_m,...$ with
corresponding magnitudes $p(x_1), p(x_2),...,p(x_m),...$. ●

Since F is monotone nondecreasing, we have

$$p(x_i) > 0 \quad , \quad \text{all } i \qquad (3.3\text{-}1)$$

and

$$\sum_i p(x_i) \leq 1 \qquad (3.3\text{-}2)$$

It is now possible to state that F may be written as the sum

$$F = \alpha F_c + (1-\alpha)F_d \qquad (3.3\text{-}3)$$

where $0 \leq \alpha \leq 1$, F_c is a distribution function that is everywhere continuous, and F_d is a discrete distribution function; that is, there exists a denumerable sequence of distinct numbers $\{x_i\}$ such that

$$\sum_i P\{X = x_i\} = 1 \qquad (3.3\text{-}4)$$

and

$$F_d(x) = P\{X \leq x\} = \sum_{x_i \leq x} P\{X = x_i\} \qquad (3.3\text{-}5)$$

Equation (3.3-3) is a simplified form of the *Lebesgue decomposition theorem*.

Proof: If F contains no jumps, the coefficient $\alpha = 1$ and $F = F_c$. If there are one or more jumps $p(x_1), p(x_2), \dots$ at $x_1, x_2, \dots$, then define

$$G_d(x) = \sum_{x_i \leq x} p(x_i) > 0 \qquad (3.3\text{-}6)$$

It is clear that F_d is now defined by

$$(1-\alpha)F_d = G_d \qquad (3.3\text{-}7)$$

where α is determined from the relationship

$$F_d(-\infty) = 0 \quad , \quad F_d(\infty) = 1 \qquad (3.3\text{-}8)$$

$$\alpha = 1 - G_d(\infty) \qquad (3.3\text{-}9)$$

It is easy to show that F_d is a discrete distribution function defined by Eq. (3.3-5). Now F_c is defined by Eq. (3.3-3); that is,

$$\alpha F_c = F - (1-\alpha)F_d \qquad (3.3\text{-}10)$$

Again it is easy to show that F_c is a continuous distribution function since $(1-\alpha)F_d$ has subtracted the jumps without otherwise affecting F. The reader may be interested in constructing a more formal proof. •

A distribution where $\alpha = 0$ is *discrete* and a distribution where $\alpha = 1$ is *continuous* ; otherwise the distribution will be called *mixed*.

3.4 *Discrete Distributions and Densities* - If X is a discrete random variable, then its distribution function F_X is a *step* function with a finite or denumerably infinite number of jumps. Thus F_X is completely characterized by the location $x_1, x_2, \dots$ of the jumps and by their magnitudes $p(x_1), p(x_2), \dots$ where $p(x_i) = P\{X = x_i\}$. The value of F_X at a jump is the value at the top of the jump since F_X is continuous from the right. It is clear from the previous discussion that

$$p(x_i) > 0, \quad i = 1,2,... \tag{3.4-1}$$

$$\sum_i p(x_i) = 1 \tag{3.4-2}$$

and that the distribution function is given by

$$F_X(x) = \sum_{x_i \leq x} p(x_i) \tag{3.4-3}$$

It will be convenient to define a discrete *(probability) density function* $f_X(x)$ by

$$f_X(x) = \begin{cases} p(x_i) & , \quad i = 1,2,... \\ 0 & , \quad \text{where } F_X \text{ contains no jumps} \end{cases} \tag{3.4-4}$$

It follows, then, that

$$f_X(x) \geq 0 \tag{3.4-5}$$

and that

$$\sum_i f_X(x_i) = 1 \tag{3.4-6}$$

Any function $f_X(x)$ which satisfies these last two equations will be called a discrete density function. We see that the probabilistic behavior of a discrete random variable is determined completely by its density function.

The Riemann-Stieltjes integral is discussed briefly in Appendix A. This integral allows us to write a discrete distribution function in the form

$$F_X(x) = \int_{-\infty}^{x} dF_X(x) = \sum_{x_i \leq x} f_X(x_i) \tag{3.4-7}$$

On the other hand, the unit step function $u(x)$ and the Dirac delta-function $\delta(x)$ of Appendix B may be used to form

$$F_X(x) = \sum_{x_i \leq x} f_X(x_i) u(x - x_i) \tag{3.4-8}$$

The formal derivative of this last expression is

$$\frac{dF_X(x)}{dx} = f_X(x_i)\delta(x - x_i) \quad , \quad -\infty < x < \infty \tag{3.4-9}$$

The principal usefulness of the approaches of this paragraph is in handling mixed distributions, which will be discussed later.

Let us now consider a number of examples of discrete distributions and densities.

Example 3.3 - The Binomial Distribution:

Define a function $b(k;n,p)$ as

$$b(k;n,p) = \begin{pmatrix} n \\ k \end{pmatrix} p^k (1-p)^{n-k}$$

for $k=0,1,...,n$ and for $0<p<1$. It is apparent that $b(k;n,p)$ is non-negative; also

$$\sum_{k=0}^{n} b(k;n,p) = [p+(1-p)]^n = 1$$

from the binomial expansion. Thus $b(k;n,p)$ is a discrete density and is called the *binomial distribution*.

Consider the particular case where $n=2$ and $p=1/2$ so that

$$b(k;2,1/2) = \frac{2!}{(2-k)!k!} \left(\frac{1}{2}\right)^n, \qquad k=0,1,2$$

This density function and the corresponding distribution function are plotted in Fig. 3.2.

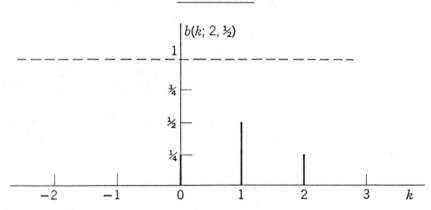

(a) The binomial density function $b(k;2,1/2)$.

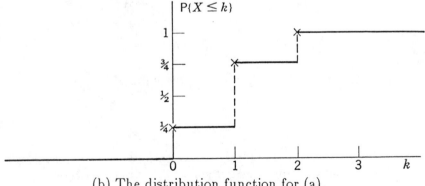

(b) The distribution function for (a).

Fig. 3.2 - A special case of the binomial distribution

Example 3.4 - The Poisson Distribution:

Define the function $p(k,\lambda)$ as

$$p(k,\lambda) = \frac{\lambda^k}{k!} e^{-\lambda} \quad , \quad \lambda > 0, \quad k=0,1,2,...$$

This function is nonnegative and sums to unity since

$$\sum_{k=0}^{\infty} \frac{\lambda^k}{k!} e^{-\lambda} = e^{\lambda} e^{-\lambda} = 1$$

Example 3.5 - The Discrete Uniform Distribution:

The distribution has jumps at $x_1, x_2, ..., x_N$ each of magnitude $1/N$; hence its density function is

$$f_X(x) = \begin{cases} 1/N & , \quad x = x_1, x_2, ..., x_N \\ 0 & , \quad \text{elsewhere} \end{cases}$$

The density and distribution function for this distribution are plotted in Fig. 2.3 for the special case where $N=2$.

3.5 *Continuous Distributions and Densities* - If X is a continuous random variable, then F_X is everywhere differentiable and

$$F_X(x) = \int_{-\infty}^{x} dF_X(y) = \int_{-\infty}^{x} f_X(y) dy \qquad (3.5\text{-}1)$$

where the *density function* $f_x(x)$ is given by

$$f_X(x) = \frac{dF_X(x)}{dx} \geq 0 \quad , \text{ for all } x \qquad (3.5\text{-}2)$$

The density function is nonnegative since F_X is monotone nondecreasing. We also have

$$\int_{-\infty}^{\infty} f_X(x) dx = \int_{-\infty}^{\infty} dF_X(x) = F_X(\infty) - F_X(-\infty) = 1 \qquad (3.5\text{-}3)$$

Any function* $f_X(x)$ which is nonnegative and satisfies

*Actually $f_X(x)$ must be a proper function. Thus we must exclude such "functions" as delta-functions which become infinite in such a way that the area under a point is non-zero. However, the restriction $f_X(x) \geq 0$ must only hold "almost everywhere" on the real line. This means that the condition need not hold on a set of points which have been assigned a probability of zero. Thus the function $f_X(x)$ may be defined arbitrarily or may remain undefined at an arbitrary set of isolated points. None of these points are of consequence in practical applications.

$$\int\limits_{-\infty}^{\infty} f_X(x)dx = 1 \tag{3.5-4}$$

will be called a continuous density function. It defines a distribution function through Eq. (3.5-1), and hence it completely characterizes the probabilistic behavior of a random variable.

We now consider a number of examples of continuous distributions.

Example 3.6 - The Uniform or Rectangular Distribution:

This distribution has been discussed in Example 2.5. Its density and distribution functions are illustrated in Fig. 2.4. Note that

$$\int\limits_{-\infty}^{\infty} f_X(x)dx = \int\limits_{a}^{b} \frac{1}{b-a} \, dx = 1$$

and

$$F_X(x) = \int\limits_{-\infty}^{x} f_X(x)dx = \begin{cases} 0 & , \ x < a \\ (x-a)/(b-a) & , \ a \le x \le b \\ 1 & , \ x > b \end{cases}$$

Example 3.7 - Exponential Distribution:

Consider the density function

$$f_X(x) = \begin{cases} e^x & , \ x \le 0 \\ 0 & , \ x > 0 \end{cases}$$

It is easy to see that this function is nonnegative and possesses unit area. The corresponding distribution function for this *negative exponential distribution* is

$$F_X(x) = \int\limits_{-\infty}^{x} f_X(y)dy = \begin{cases} e^x & , \ x \le 0 \\ 1 & , \ x > 0 \end{cases}$$

Both $f_X(x)$ and $F_X(x)$ are illustrated in Fig. 3.3.

The *double-exponential distribution* has a density function $f_X(x)$ given by

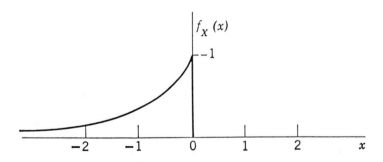

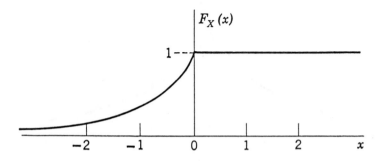

Fig. 3.3 - The density function and the distribution function
for the negative exponential distribution.

$$f_X(x) = \frac{\alpha}{2} e^{-\alpha|x|} \, , \alpha > 0 \, , \quad -\infty < x < \infty$$

Again it is easy to check that this function is a density function.
The corresponding distribution function is

$$F_X(x) = \begin{cases} \dfrac{1}{2} \, e^{\alpha x} & , \ x \leq 0 \\ 1 - \dfrac{1}{2} \, e^{-\alpha x} & , \ x > 0 \end{cases}$$

Both $f_X(x)$ and $F_X(x)$ are plotted in Fig. 3.4. Note that the
negative exponential distribution could have been defined more
generally by including the parameter α. In the same way, a *positive exponential* distribution may be developed.

Example 3.8 - The Cauchy Distribution:
    This distribution has a density function given by

$$f_X(x) = \frac{1}{\pi(1+x^2)} \quad , \quad -\infty < x < \infty$$

The distribution function is easily found to be

$$F_X(x) = \int_{-\infty}^{x} \frac{dy}{\pi(1+y^2)} = \frac{1}{\pi}\,[\text{arc tan } x + \frac{\pi}{2}]$$

It is clear from this last expression that

$$F_X(\infty) = 1$$

and, hence, that $f_X(x)$ is a density function.

___

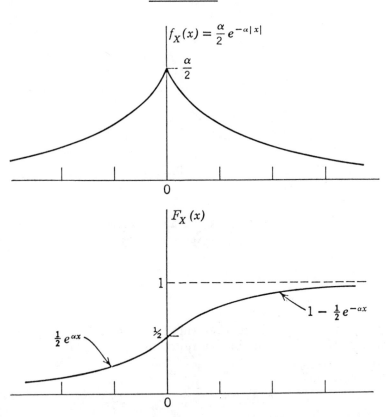

Fig. 3.4 - The density function and the distribution function for the double-exponential distribution.

Example 3.9 - The Unit Normal Distribution

This distribution is the most important of the continuous distributions and will be studied in later chapters in considerable detail. It will be denoted by N(0,1). Its density function is

$$\phi_X(x) = \frac{1}{\sqrt{2\pi}} \ e^{-x^2/2} \ , \quad -\infty < x < \infty$$

and its distribution function is

$$\Phi_X(x) = \frac{1}{\sqrt{2\pi}} \int\limits_{-\infty}^{x} e^{-y^2/2} \ dy$$

It is clear that $\phi_X(x)$ is nonnegative. That it possesses unit area may be shown by considering the product

$$J^2 = \frac{1}{\sqrt{2\pi}} \int\limits_{-\infty}^{\infty} e^{-x^2/2} \ dx \ \frac{1}{\sqrt{2\pi}} \int\limits_{-\infty}^{\infty} e^{-y^2/2} \ dy$$

or

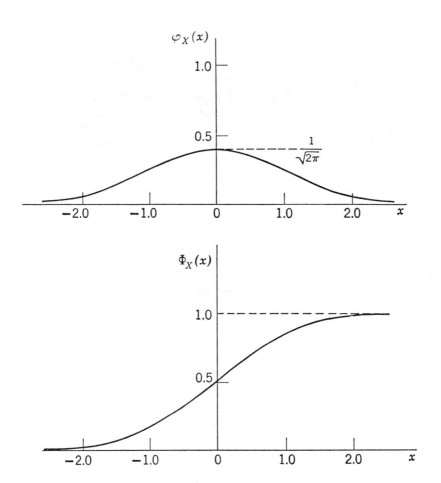

Fig. 3.5 - The density function and the distribution function for the unit normal distribution.

$$J^2 = \frac{1}{2\pi} \int\limits_{-\infty}^{\infty} \int\limits_{-\infty}^{\infty} e^{-(x^2+y^2)/2} \, dx \; dy$$

A change to polar coordinates with $dx \; dy = \nu d\nu d\theta$ yields

$$J^2 = \frac{1}{2\pi} \int\limits_{0}^{\infty} \nu \, e^{-\nu^2/2} \{ \int\limits_{0}^{2\pi} d\theta \} d\upsilon \equiv \int\limits_{0}^{\infty} \nu e^{-\nu^2/2} \, d\nu = -e^{-\nu^2/2} \Big|_{0}^{\infty} = 1$$

Since $J^2 = 1$ and $J > 0$, it is clear that $J = 1$ and that $\phi_X(x)$ is a density function.

A plot of $\phi_X(x)$ and of $\Phi_X(x)$ is shown in Fig. 3.5.

It is obvious from the preceding examples that continuous densities and distributions are easy to construct. One simply chooses a nonnegative function and normalizes it to unit area to produce a density function. The associated distribution function is obtained then by integration. From a practical point of view, the question of interest is whether or not random variables with these particular density and distribution functions correspond to underlying random experiments that can be used as a model of reality. In other words, do random variables with these density and distribution functions occur in practice? It turns out, that each of the distributions discussed in these examples is useful in many practical applications.

3.6 *Mixed Distributions and Densities* - The case may arise where both components of Eq. (3.3-3) are present; that is, the distribution may contain both a continuous and a discrete component. Such distributions will be called *mixed* and do occur in a number of practical applications. In this case, α of Eq. (3.3-3) is neither zero nor unity, and it is convenient to rewrite the equation as

$$F_X(x) = \alpha \int\limits_{-\infty}^{x} dF_{Xc}(y) + (1-\alpha) \int\limits_{-\infty}^{x} dF_{Xd}(y) \qquad (3.6\text{-}1)$$

or

$$F_X(x) = \alpha \int\limits_{-\infty}^{x} f_{Xc}(y) \, dy + (1-\alpha) \sum_{x_i \leq x} f_{Xd}(x_i) u(x - x_i) \quad (3.6\text{-}2)$$

where f_{Xc} is a continuous density given by

$$f_{Xc}(x) = \frac{dF_{Xc}(x)}{dx} \; , \qquad (3.6\text{-}3)$$

the set $\{x_i\}$ are the locations of jumps of size $f_{Xd}(x_i)$, and $u(x)$ is the unit step-function of Appendix B.

Example 3.10 - A Mixed Distribution

Consider the random variable X with

$$P\{X=1/8\} = 1/4 \quad , \quad P\{X=1\} = 1/8$$

and with the probability of 5/8 of being uniformly distributed in $(\frac{1}{8}, 1)$. The distribution function and its decomposition is shown in Fig. 3.6.

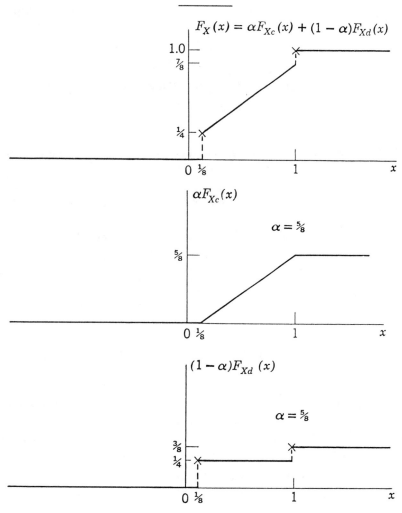

Fig. 3.6 - A mixed distribution and its decomposition.

3.7 Further Properties and Comments - Every random variable X possesses a distribution function F_X defined by Eq. (3.1-1); conversely, for every function F with Properties (1), (2), and (3) developed in Section 3.2, there is a random variable for which F is the distribution function, as shown by the following:

Converse Property:

If F is a function defined on $(-\infty,\infty)$ and if it satisfies Properties (1), (2), and (3) of Section 3.2, then there exists a probability space $\{S,\mathbf{B},P\}$ and a random variable X defined on S such that $F_X(x) = F(x)$ for all real x.

Proof: Let the sample space S be the real line $(-\infty,\infty)$, and let $\mathbf{B}$ be the sigma field generated by the subintervals of $(-\infty,\infty)$. [The members of this field are called the *Borel subsets* of $(-\infty,\infty)$.] The probability measure P will be determined from the relationship

$$P\{x_1 < X \leq x_2\} = \int_{x_1}^{x_2} dF(x) \tag{3.7-1}$$

where $dF(x)$ is the probability differential*. Now define X by

$$X(\omega) = \omega \quad \text{for all} \quad \omega \in S \tag{3.7-2}$$

and it is clear that $P\{X \leq x\} = F(x)$. •

The *convolution* of two distribution functions F and G will be denoted by $F*G$ and will be defined as

$$F*G(x) = \int_{-\infty}^{\infty} F(x-y)\, dG(y) \ , \quad -\infty < x < \infty \tag{3.7-3}$$

where the integral is the Reimann-Stieltjes integral discussed in Appendix A. It can be shown that $F*G$ is a distribution function. In view of the previous paragraph it is only necessary to show that $F*G$ satisfies the three properties given in Section 3.2.

Proof: Property (1) states that $F*G$ is monotone nondecreasing. Form the relationship

$$F*G(x+h) - F*G(x) = \int_{-\infty}^{\infty} [F(x+h-y)-F(x-y)]\, dG(y) \tag{3.7-4}$$

where $h > 0$. Since $F(x+h-y) \geq F(x-y)$, the integrand is nonnegative; hence $F*G(x+h) \geq F*G(x)$ and $F*G$ is monotone nondecreasing. Property (2) states that $F*G$ is right-continuous. Form the limit

$$\lim_{\substack{h \to 0 \\ h > 0}} [F*G(x+h) - F*G(x)] =$$

$$\tag{3.7-5}$$

$$= \int_{-\infty}^{\infty} \lim_{\substack{h \to 0 \\ h > 0}} [F(x+h-y)-F(x-y)]\, dG(y)$$

---

*A more rigorous statement would be to say that P is the Lebesgue-Stieltjes measure determined by F over $\mathbf{B}$.

The right side of this expression is zero since $F(x)$ is right continuous. The operations of limit and integration may be interchanged because the integral is absolutely convergent. Property (3) states that $F*G(-\infty) = 0$ and $F*G(\infty) = 1$. As before

$$\lim_{x \to -\infty} F*G(x) = \int_{-\infty}^{\infty} [\lim_{x \to -\infty} F(x-y)] dG(y) = 0 \qquad (3.7\text{-}6)$$

and

$$\lim_{x \to \infty} F*G(x) = \int_{-\infty}^{\infty} [\lim_{x \to \infty} F(x-y)] dG(y) = 1 \quad \bullet \qquad (3.7\text{-}7)$$

Convolutions of distribution functions will arise later when sums of independent random variables are considered.

The discussion in this chapter began with the definition of a distribution function and the density function arose as a consequence of this definition. Another approach would be to start with the density function. In this approach we begin with the discrete set of values $\{x_i\}$; the associated discrete density function $f_X(x)$ is defined to be a real-valued function with the two properties:

$$f_X(x) = \begin{cases} >0 , & \text{for } x = x_i \\ =0 , & \text{for } x \neq x_i \end{cases} \qquad (3.7\text{-}8)$$

and

$$\sum_i f_X(x_i) = 1 \qquad (3.7\text{-}9)$$

A function F_X is now defined by

$$F_X(x) = \sum_{x \leq x_i} f(x_i) \qquad (3.7\text{-}10)$$

This function is shown to be a distribution function by showing that it possesses Properties (1), (2), and (3) of Section 3.2. The proof is left to the reader.

In the same way, in the continuous case, a continuous density function $f_X(x)$ is defined as a real-valued function with the two properties:

$$f_X(x) \geq 0 \qquad (3.7\text{-}11)$$

and

$$\int_{-\infty}^{\infty} f_X(x) \, dx = 1 \qquad (3.7\text{-}12)$$

Again a function $F_X(x)$ is defined by

$$F_X(x) = \int_{-\infty}^{x} f_X(y) \, dy \qquad (3.7\text{-}13)$$

and this function is shown to be a distribution function. As a matter of interest, we outline the proof in this case.

Property (1) - $F_X(x)$ *is monotone nondecreasing*
Proof:

$$F_X(x+h) = F_X(x) + \int_x^{x+h} f_X(y)dy \quad , \quad h > 0$$

Since $f_X(x)$ is nonnegative, the last term must be nonnegative and we have $F_X(x+h) \geq F_X(x)$, $h > 0$. •

Property (2) - $F_X(x)$ *is right-continuous*
Proof:

$$F_X(x+0) - F_X(x) = \lim_{\substack{h \to 0 \\ h > 0}} \int_x^{x+h} f_X(y)\, dy = \lim_{\substack{h \to 0 \\ h > 0}} h\, f_X(x) = 0 \quad •$$

Property (3) - $F_X(-\infty) = 0$; $F_X(\infty) = 1$
Proof:

$$F_X(-\infty) = \lim_{x \to -\infty} \int_{-\infty}^x f_X(y)\, dy = 0$$

$$F_X(\infty) = \lim_{x \to \infty} \int_{-\infty}^x f_X(y)\, dy = 1 \quad •$$

Thus $F_X(x)$ can be shown to be a distribution function in both the discrete and continuous cases. From the converse property developed at the beginning of this section, there is a random variable X for which F_X is the distribution function.

The *law of X* , denoted by $\mathbf{L}(X)$, is defined as the distribution of X. If the random variables X and Y have the same distribution function, it is conventional to write

$$\mathbf{L}(X) = \mathbf{L}(Y) \tag{3.7-14}$$

or

$$X \overset{d}{=} Y \tag{3.7-15}$$

This topic will be taken up again in Section 8.2 in a discussion of the convergence of sequences of random variables.

The *support* of F is the smallest closed set C such that $F(C)=1$. A *point of increase* of F is a point x_0 such that, for every neighborhood* N of x_0 ,$F(N) > 0$. The set of all points of *increase* is the support of F.

3.8 *Bivariate Distributions* - In many studies of random phenomena, more than one random variable will be involved. We will call these *multivariate* cases. The simplest to consider is the one involving two random variables, the *bivariate* case.

The *joint* or *bivariate* distribution function $F_{X,Y}(x,y)$ of the random variables X and Y is defined as

$$F_{X,Y}(x,y) = P\{X \leq x \text{ and } Y \leq y\} =$$

(3.8-1)

$$P\{X \leq x \cap Y \leq y\}$$

for $-\infty < x,y < \infty$. As in the univariate case, this function has a number of properties that follow immediately from the definition.

(1a) *For any* $h \geq 0$ *and* $k \geq 0$, *the relationship holds that*

$$F_{X,Y}(x+h,y+k) - F_{X,Y}(x+h,y) -$$

(3.8-2)

$$F_{X,Y}(x,y+k) + F_{X,Y}(x,y) \geq 0$$

This last expression and Property (1b) are analogous to Property (1) of Section 3.2 which states that the univariate distribution function $F_X(x)$ is monotone nondecreasing. The proof there consisted of showing that $F_X(x+h)$ was the probability of an event containing the event with probability $F_X(x)$. In this case the proof proceeds in the same way.

Proof: Define the events A, B, C, and D by

$$A - \{X \leq x+h\}$$
$$B = \{Y \leq y+k\}$$
$$C = \{X \leq x\}$$
$$D = \{Y \leq y\}$$

Then it follows that $[C \cap B] \subset [A \cap B]$ and, as a consequence,

$$P\{A \cap B\} - P\{C \cap B\} = P\{[A \cap B] - [C \cap B]\} \geq 0$$

Also $[C \cap D] \subset [A \cap D]$ so that

$$P\{A \cap D\} - P\{C \cap D\} = P\{[A \cap D] - [C \cap D]\} \geq 0$$

---

*Let x_0 be a given point and let $h > 0$. The open interval (x_0-h, x_0+h) is called a *neighborhood* with x_0 as center and of radius h. This neighborhood of x_0 is denoted by $N(x_0;h)$, by $N(x_0)$, or simply by N.

In the same way, $[A \cap D]-[C \cap D] \subset [A \cap B]-[C \cap B]$ so that

$$P\{[A \cap B]-[A \cap D]-[C \cap B]+[C \cap D]\} \geq 0$$

But this last expression is just Eq. (3.8-2). •

Understanding this proof may be improved by consulting Fig. 3.7 which shows the different sets.

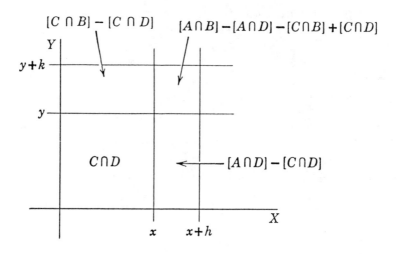

Fig. 3.7 - The region $A \cap B$ and its subdivisions.

(1b) *For any $h \geq 0$ and $k \geq 0$, the relationships hold that*

$$F_{X,Y}(x+h,y) \geq F_{X,Y}(x,y) \tag{3.8-3}$$

$$F_{X,Y}(x,y+k) \geq F_{X,Y}(x,y) \tag{3.8-4}$$

Proof: The proof has essentially been given. Since $[C \cap D] \subset [A \cap D]$ and $[C \cap D] \subset [C \cap B]$, it follows that $P\{A \cap D\} \geq \} P\{C \cap D\}$ and that $P\{C \cap B\} \geq P\{C \cap D\}$. •

(2) $F_{X,Y}(x,y)$ *is right-continuous in each argument; that is,*

$$\lim_{\substack{h \to 0 \\ h > 0}} F_{X,Y}(x+h,y) = F_{X,Y}(x,y) \text{ for all } x,y \tag{3.8-5}$$

$$\lim_{\substack{k \to 0 \\ k > 0}} F_{X,Y}(x,y+k) = F_{X,Y}(x,y) \text{ for all } x,y \tag{3.8-6}$$

The proof is the same as for Property (2) in Section 3.2 and will not be repeated.

(3)

$$\lim_{x \to -\infty} F_{X,Y}(x,y) = \lim_{y \to -\infty} F_{X,Y}(x,y) = 0,$$

$$(3.8\text{-}7)$$

$$\lim_{\substack{x \to \infty \\ y \to \infty}} F_{X,Y}(x,y) = 1$$

Again the proof is the same as in Section 3.2.

As in the univariate case, not only do all bivariate distribution functions possess the three properties just developed, but to any function F possessing these properties, there corresponds a probability space $(S,\mathbf{B},P)$ and two random variables X and Y defined on S such that $F_{X,Y} = F$. This converse property was developed in Section 3.7 for the univariate case. For further details, see Cramer (1946).

The three properties just developed for $F_{X,Y}$ are somewhat more complicated than in the univariate case. In particular, Property (1a) has no one-dimensional counterpart. It is necessary, however, to insure that all probabilities involving the random variables X and Y be nonnegative. For example the probability that X and Y lie in some rectangle in the X–Y plane is given by

$$P\{a < X \leq b \text{ and } c < Y \leq d\}$$

$$(3.8\text{-}8)$$

$$= F_{X,Y}(b,d) - F_{X,Y}(a,d) - F_{X,Y}(b,c) + F_{X,Y}(a,c)$$

and not by $F_{X,Y}(b,d) - F_{X,Y}(a,c)$. It is easy to construct a function F which satisfies Properties (1b), (2), and (3) but violates Property (1a). Such a function cannot be a probability and, hence, a distribution function.

Example 3.11

Let a function $F(x,y)$ be defined by

$$F(x,y) = \begin{cases} 0 & \text{if } x \leq 0 \text{ or } y \leq 0 \text{ or } x+y \leq 1 \\ 1 & \text{elsewhere} \end{cases}$$

This function is nondecreasing in each argument [Property (1b)], is right-continuous [Property (2)], and has the correct behavior at $\pm\infty$ [Property (3)]. Nevertheless, it does not possess Property (1a) and is not a distribution function. Let $a = c = 1/2$ and $b = d = 1$ and substitute into Eq. (3.8-8); we obtain

$$F(1,1) - F(1,1/2) - F(1/2,1) + (1/2,1/2) = 1-1-1+0 = -1$$

In addition to the properties already discussed, the bivariate distribution function must obey certain *consistency* relationships in the sense that legitimate univariate distribution functions for X and Y must be derivable from $F_{X,Y}$. It is clear that the following must hold:

$$P\{X \leq x, \text{ no condition on } Y\} = F_X(x) \qquad (3.8\text{-}9)$$

and

$$P\{\text{no condition on } X, Y \leq y\} = F_Y(y) \qquad (3.8\text{-}10)$$

where $F_X(x)$ and $F_Y(y)$ are the univariate distribution functions of X and Y, respectively. However, these last two equations can be written as

$$F_{X,Y}(x,\infty) = F_X(x) \qquad (3.8\text{-}11)$$

$$F_{X,Y}(\infty,y) = F_Y(y) \qquad (3.8\text{-}12)$$

since the set $\{X \leq x\}$ and the set $\{Y \leq y\}$ both approach S as x and y approach infinity, respectively. The functions F_X and F_Y are often called the *marginal distributions* of $F_{X,Y}$.

Example 3.12 - The Bivariate Normal Distribution

The most common bivariate distribution that will be encountered later is the bivariate normal distribution given in normalized form by

$$\Phi_{X,Y}(x,y) = K \int\limits_{-\infty}^{y} \int\limits_{-\infty}^{x} e^{-Q(u,v)} \, du \, dv$$

where K is a constant given by

$$K = \frac{1}{2\pi\sqrt{1-\rho^2}}$$

$Q(u,v)$ is a quadratic form given by

$$Q(u,v) = \frac{u^2 - 2\rho uv + v^2}{2(1-\rho^2)}$$

and where ρ is a real number called the *correlation coefficient* satisfying the inequality

$$|\rho| < 1$$

It is easily shown that $\Phi_{X,Y}$ is a bivariate distribution function.

Consider the limiting case where $\rho = 0$; then $\Phi_{X,Y}(x,y)$ can be written as

$$\Phi_{X,Y}(x,y) = \frac{1}{\sqrt{2\pi}} \int_{-\infty}^{x} e^{-u^2/2} \, du \; \frac{1}{\sqrt{2\pi}} \int_{-\infty}^{y} e^{-v^2/2} \, dv$$

$$= \Phi_X(x)\Phi_Y(y)$$

where $\Phi_X(x)$ and $\Phi_Y(y)$ are the distribution functions for the univariate unit normal distributions of Example 3.9. In this case the bivariate distribution function factors into the product of the two marginal distribution functions. As will be discussed later, this factorization is a necessary and sufficient condition that X and Y be independent.

In the general case where $0 < |\rho| < 1$, the marginal distributions are easily found. Since $\Phi_{X,Y}$ is symmetrical in x and y, only one need be determined. We rewrite $\Phi_{X,Y}$ as

$$\Phi_{X,Y}(\infty,y) = \Phi_Y(y) = K \int_{-\infty}^{y} e^{-v^2/2} \int_{-\infty}^{\infty} e^{-\frac{(u-\rho v)^2}{2(1-\rho^2)}} \, du \; dv$$

In the inner integral, make the linear change in variable $w = (u-\rho v)/\sqrt{1-\rho^2}$ to obtain

$$\sqrt{1-\rho^2} \int_{-\infty}^{\infty} e^{-w^2/2} \, dw = \sqrt{2\pi} \sqrt{1-\rho^2}$$

This last result was shown in Example 3.9. Now Φ_Y may be written as

$$\Phi_Y(y) = \frac{1}{\sqrt{2\pi}} \int_{-\infty}^{y} e^{-v^2/2} \, dv$$

which is the univariate unit normal distribution function.

Note that the expression for $\Phi_{X,Y}$ becomes meaningless for $\rho = 1$; in this case X and Y are linearly related.

3.9 *Bivariate Density Functions* - The bivariate distribution function $F_{X,Y}(x,y)$ is called continuous when a function $f_{X,Y}(u,v)$ exists such that

$$F_{X,Y}(x,y) = \int_{-\infty}^{y} \int_{-\infty}^{x} f_{X,Y}(u,v) \, du \; dv \qquad (3.9\text{-}1)$$

for all x and y. The function $f_{X,Y}$ is called the *joint or bivariate density function* for the continuous random variables X and Y. As in the univariate case, Eq. (3.9-1) is equivalent, from a practical point of view, to the relationship

$$f_{X,Y}(x,y) = \frac{\partial}{\partial x}\frac{\partial}{\partial y} F_{X,Y}(x,y) \qquad (3.9\text{-}2)$$

Equivalently, a bivariate density function for the continuous random variables X and Y is a function $f(x,y)$ satisfying the two properties:

$$f(x,y) \geq 0 \qquad (3.9\text{-}3)$$

$$\int\limits_{-\infty}^{\infty}\int\limits_{-\infty}^{\infty} f(x,y)\ dx\ dy = 1 \qquad (3.9\text{-}4)$$

It is clear that the bivariate density function $f_{X,Y}$ satisfies the same type of consistency relationship as its distribution function. We write

$$F_Y(y) = F_{X,Y}(\infty,y) = \int\limits_{-\infty}^{y}\left[\int\limits_{-\infty}^{\infty} f_{X,Y}(u,v)\ du\right] dv$$

$$= \int\limits_{-\infty}^{y} f_Y(v)\,dv \qquad (3.9\text{-}5)$$

Thus the univariate density f_Y is related to $f_{X,Y}$ by

$$f_Y(y) = \int\limits_{-\infty}^{\infty} f_{X,Y}(u,y)\ du \qquad (3.9\text{-}6)$$

and, in the same way,

$$f_X(x) = \int\limits_{-\infty}^{\infty} f_{X,Y}(x,v)\ dv \qquad (3.9\text{-}7)$$

In the context of the last two equations, the function f_X is sometimes called the *marginal density* of X and f_Y is called the *marginal density* of Y. It should be mentioned in passing that both f_X and f_Y are easily obtained from $f_{X,Y}$ but, in general, $f_{X,Y}$ cannot be found from a knowledge of *either or both* f_X and f_Y. The same remark is true of F_X, F_Y and $F_{X,Y}$.

The bivariate distribution function $F_{X,Y}(x,y)$ is called *discrete* if there is a denumerable set of distinct points $(x_1,y_1),(x_1,y_2),(x_2,y_1),(x_2,y_2)$,...,(x_i,y_j),...with a set of associated positive numbers $p(x_1,y_1)$, $p(x_1,y_2)$, $p(x_2,y_1)$, $p(x_2,y_2),...,p(x_i,y_j)$,...such that

$$F_{X,Y}(x,y) = \sum_{x_i \leq x}\sum_{y_j \leq y} p(x_i,y_j) \qquad (3.9\text{-}8)$$

and, consequently,

$$p\left(x_i,y_j\right) = P\left\{X=x_i \text{ and } Y=y_j\right\} \qquad (3.9\text{-}9)$$

As in the univariate case, it is convenient to define a discrete bivariate density function by

$$f_{X,Y}(x,y) = \begin{cases} p\left(x_i,y_j\right), & \text{for } x=x_i \text{ and } y=y_j \\ 0, & \text{otherwise} \end{cases} \qquad (3.9\text{-}10)$$

The marginal densities f_X and f_Y are given by

$$f_X(x) = \sum_{y=y_j} f_{X,Y}(x,y) \qquad (3.9\text{-}11)$$

$$f_Y(y) = \sum_{x=x_i} f_{X,Y}(x,y) \qquad (3.9\text{-}12)$$

Note that a discrete bivariate density $f_{X,Y}$ is defined by

$$f_{X,Y}(x,y) = \begin{cases} f_{X,Y}(x_i,y_j) \geq 0, & \text{for } x=x_i \text{ and } y=y_j \\ 0, & \text{otherwise} \end{cases} \qquad (3.9\text{-}13)$$

and

$$\sum_{i=1}^{\infty} \sum_{j=1}^{\infty} f_{X,Y}(x_i,y_j) = 1 \qquad (3.9\text{-}14)$$

3.10 *Multivariate Distributions* - When more than two random variables are involved, the resulting distributions are usually called *multivariate*. Consider the random vector $\mathbf{X}_n = (X_1,X_2,...,X_n)$. The joint distribution function or the multivariate distribution function of the X_i is defined as

$$F_{\mathbf{X}_n}(\mathbf{x}_n) = P\left\{\bigcap_{i=1}^{n} [X_i \leq x_i]\right\} \qquad (3.10\text{-}1)$$

for $-\infty < x_i < \infty$ and for $1 \leq i \leq n$ where $\mathbf{x}_n$ is the real vector $\mathbf{x}_n = (x_1,x_2,...,x_n)$.

As in the bivariate case, this function must obey certain consistency relationships. Specifically, for $n \geq 2$, it must follow that

$$\lim_{x_n \to \infty} F_{\mathbf{X}_n}(\mathbf{x}_n) = F_{\mathbf{X}_{n-1}}(\mathbf{x}_{n-1}) \qquad (3.10\text{-}2)$$

Proof: As $x_n \to \infty$, the set $[X_n \leq x_n] \to S$; hence

$$\lim_{x_n \to \infty} F_{\mathbf{X}_n} (\mathbf{x}_n) = P \{ \bigcap_{i=1}^{n-1} [X_i \le x_i] \} \quad \bullet$$

The function $F_{\mathbf{X}_{n-1}}$ is called the *marginal distribution* of $F_{\mathbf{X}_n}$. More generally, the joint distribution function of any subset of $\mathbf{X}_n$ is often called a marginal distribution of $F_{\mathbf{X}_n}$.

The multivariate distribution function $F_{\mathbf{X}_n}$ possesses the same set of three fundamental properties as in the bivariate case. As a matter of convenience, we introduce some additional notation. Define the real vectors $\mathbf{a}_n$ and $\mathbf{b}_n$ where

$$\mathbf{a}_n = (a_1, a_2, ..., a_n)$$
$$\mathbf{b}_n = (b_1, b_2, ..., b_n)$$

A *cell* in R_n will be denoted by $(\mathbf{a}_n, \mathbf{b}_n)$ and will be defined by the set

$$(\mathbf{a}_n, \mathbf{b}_n) = \{ \mathbf{x}_n \mid a_i < x_i \le b_i, \ 1 \le i \le n \} \tag{3.10-3}$$

Each cell possesses 2^n vertices v_j in the space R_n. Let the set $\{Z_n^k\}$ be the set of all n-tuples each containing exactly k elements a_i and $n-k$ elements b_i; that is, each Z_n^k is given by

$$Z_n^k = (z_1, z_2, ..., z_n)$$

where k of the z_i are a_i' s. There are $\begin{pmatrix} n \\ k \end{pmatrix}$ elements in the set $\{Z_n^k\}$. Now the set

$$Z = \bigcup_{k=0}^{n} \{Z_n^k\}$$

is the set of 2^n vertices v_j of the cell. For example, in the bivariate case, there are $2^2 = 4$ vertices. The sets $\{Z_2^k\}$ are

$$\{Z_2^0\} = \{(b_1, b_2)\}$$
$$\{Z_2^1\} = \{(a_1, b_2)(b_1, a_2)\}$$
$$\{Z_2^2\} = \{(a_1, a_2)\}$$

and the set Z is

$$Z = \{(a_1, a_2), (a_1, b_2), (b_1, a_2), (b_1, b_2)\}$$

Each of the four elements of Z is one of the vertices of the cell $(\mathbf{a}_2, \mathbf{b}_2)$. Now the three fundamental properties of $F_{\mathbf{X}_n}$ may be written as

(1a) *For every cell* $(\mathbf{a}_n, \mathbf{b}_n)$ *in* R_n , *the relationship holds that*

$$\sum_{k=0}^{n} (-1)^k \sum_{v_i \in \{Z_n^k\}} F_{\mathbf{X}_n} (v_i) \ge 0 \tag{3.10-4}$$

We shall not prove this result although a proof is given in Breiman (1968). Note that, for $n = 2$, Eq. (3.10-4) reduces to Eq. (3.8-2) with $a_1 = x$, $a_2 = x + h$, $b_1 = y$, and $b_2 = y + k$.

(1b) *For any* $h_i \geq 0, 1 \leq i \leq n$, *and for all* x_i *the relationship holds that*

$$F_{X_n}(x_1, ..., x_i + h_i, ..., x_n) \geq F_{X_n}(x_1, ..., x_i, ..., x_n) \qquad (3.10\text{-}5)$$

This proof that F_{X_n} is monotone nondecreasing in each argument has already been given in the bivariate case.

(2) $F_{X_n}(\mathbf{x}_n)$ *is right-continuous in each argument;* that is, for any $1 \leq i \leq n$ and for all x_i,

$$\lim_{\substack{h_i \to 0 \\ h_i > 0}} F_{X_n}(x_1, ..., x_i + h_i, ..., x_n) = F_{X_n}(\mathbf{x}_n) \qquad (3.10\text{-}6)$$

The proof is the same as in the univariate case.

(3a) *For each* $1 \leq i \leq n$,

$$\lim_{x_i \to -\infty} F_{X_n}(\mathbf{x}_n) = 0 \qquad (3.10\text{-}7)$$

(3b) *and,*

$$\lim_{\substack{\min x_i \to \infty \\ 1 \leq i \leq n}} F_{X_n}(\mathbf{x}_n) = 1 \qquad (3.10\text{-}8)$$

These Properties (3a) and (3b) are proved exactly as in Section 3.2 for the univariate case.

As in Section 3.7 it can be shown that, for any function F with these three properties, there exists a probability space $(X, \mathbf{B}, P)$ and a random vector $\mathbf{X}_n$ defined on S such that $F = F_{X_n}$. Thus, as before, these properties are necessary and sufficient properties for a distribution function.

A random vector X_n is said to be *continuous* if there exists a continuous multivariate function $f_{X_n}(\mathbf{x}_n)$, such that

$$F_{X_n}(\mathbf{x}_n) = \int_{-\infty}^{x_1} \cdots \int_{-\infty}^{x_n} f_{X_n}(\mathbf{x}_n)\, dx_1 ... dx_n \qquad (3.10\text{-}9)$$

and

$$f_{X_n}(\mathbf{x}_n) = \frac{\partial}{\partial x_1} \cdots \frac{\partial}{\partial x_n} F_{X_n}(\mathbf{x}_n) \qquad (3.10\text{-}10)$$

Also marginal densities $f_{X_{n-1}}$ are given by

$$f_{X_{n-1}}(\mathbf{x}_{n-1}) = \int_{-\infty}^{\infty} f_{X_n}(\mathbf{x}_n)\, dx_n \qquad (3.10\text{-}11)$$

Similarly, for the discrete multivariate distribution function $F_{X_n}(\mathbf{x}_n)$ of a *discrete* random variable, there exists a denumerable set of positive numbers $p(x_{i1},...,x_{jn})$ such that

$$p(x_{i1},...,x_{jn}) = P\{X_1 = x_{i1} \text{ and } \cdots \text{ and } X_n = x_{jn}\} \quad (3.10\text{-}12)$$

and, consequently,

$$F_{X_n}(\mathbf{x}_n) = \sum_{x_{i1} \leq x_1} \cdots \sum_{x_{jn} \leq x_n} p(x_{i1},...,x_{jn}) \qquad (3.10\text{-}13)$$

A discrete multivariate density function is defined by

$$f_{X_n}(\mathbf{x}_n) = \begin{cases} p(x_{i1},...,x_{jn}) & , \quad i,...,j = 1,2,..., \\ \\ 0 & , \quad \text{where } F_{X_n} \text{ is non--increasing} \end{cases} \qquad (3.10\text{-}14)$$

3.11 - *Independence* - It was pointed out in Section 3.5 that random variables are independent if and only if the underlying events, on which the random variables are defined, are independent. It follows as a direct consequence that the multivariate distribution function of a set of mutually independent random variables must factor into the product of the individual univariate distributions. More specifically, let $\mathbf{X} = \{X_\lambda, \lambda \in \Lambda\}$ be a set of random variables indexed by λ. These random variables are independent if and only if

$$F_{X_{\lambda_1},...,X_{\lambda_n}}(x_1,...,x_n) = P\{\bigcap_{i=1}^{n} \{X_{\lambda_i} \leq x_i\}\}$$

$$(3.11\text{-}1)$$

$$= \prod_{i=1}^{n} F_{X_{\lambda_i}}(x_i)$$

for every positive integer n, every n distinct elements $\lambda_1,...,\lambda_n$ in the set Λ, and all $\mathbf{x} = (x_1,...,x_n) \in R_n$. The function $F_{X_{\lambda_i}}$ is the distribution function of the random variable X_{λ_i}, that is,

$$F_{X_{\lambda_i}}(x_i) = P\{X_{\lambda_i} \leq x_i\} \qquad (3.11\text{-}2)$$

In the bivariate case, Eq. (3.11-1) becomes

$$F_{X,Y}(x,y) = F_X(x)\, F_Y(y) \qquad (3.11\text{-}3)$$

For the continuous case, independence could have been treated equally well in terms of the factorization of multivariate density functions. Consider the bivariate case where X and Y are independent. We have

$$P\{a < X \leq b \text{ and } c < Y \leq d\} = P\{a < X \leq b\}P\{c < Y \leq d\}$$

$$(3.11\text{-}4)$$

$$= [F_X(b)-F_X(a)]\,[F_Y(d)-F_Y(c)]$$

In the limit, these expressions become

$$P\{x < X \leq x+dx \text{ and } y < Y \leq y+dy\} = f_{X,Y}(x,y)\,dx\,\,dy$$

$$(3.11\text{-}5)$$

$$= f_X(x)\,dx\,\,f_Y(y)\,dy$$

or, finally,

$$f_{X,Y}(x,y) = f_X(x)\,f_Y(y) \tag{3.11-6}$$

It is left to the reader to construct the equivalent of Eq. (3.11-1) using multivariate density functions.

Note that a necessary and sufficient condition for Eq. (3.11-6) to hold and, hence, for X and Y to be independent is that $f_{X,Y}$ can be written as

$$f_{X,Y}(x,y) = f(x)\,g(y) \tag{3.11-7}$$

where f is a function of x only and g is a function of y only.

Proof: The condition is obviously necessary. To prove sufficiency, we use Eq.(3.11-16) to write

$$f_X(x) = \int_{-\infty}^{\infty} f_{X,Y}(x,y)\,dy = f(x)K_1 \tag{3.11-8}$$

where K_1 is a constant given by

$$K_1 = \int_{-\infty}^{\infty} g(y)\,dy \tag{3.11-9}$$

In the same way, we obtain

$$f_Y(y) = \int_{-\infty}^{\infty} f_{X,Y}(x,y)\,dx = g(y)\,K_2 \tag{3.11-10}$$

where

$$K_2 = \int_{-\infty}^{\infty} f(x)\,dx \tag{3.11-11}$$

in other words, the product $f(x)g(y)$ must obey

$$f(x)g(y) = \frac{1}{K_1K_2}\,f_X(x)f_Y(y) \tag{3.11-12}$$

It follows from Eqs. (3.11-6) and (3.11-7) that

$$1 = \int\limits_{-\infty}^{\infty} \int\limits_{-\infty}^{\infty} f(x)g(y)dxdy$$

$$(3.11\text{-}13)$$

$$= \frac{1}{K_1 K_2} \int\limits_{-\infty}^{\infty} f_X(x)f_Y(y)dxdy = \frac{1}{K_1 K_2}$$

Hence $K_1 K_2 = 1$, $f(x) = f_X(x)$, and $g(y) = f_Y(y)$.

Example 3.13

Two statistically independent random variables X and Y have density functions

$$f_X(u) = \begin{cases} (u+1)^{-2} , & u > 0 \\ 0 , & u \leq 0 \end{cases}$$

and

$$f_Y(v) = \begin{cases} (v+1)^{-2} , & v > 0 \\ 0 , & v \leq 0 \end{cases}$$

Since X and Y are statistically independent, it follows that the joint density function $f_{X,Y}$ is just the product

$$F_{X,Y}(u,v) = \begin{cases} (u+1)^{-2}(v+1)^{-2} , & u > 0 \text{ and } v > 0 \\ 0 , & \text{elsewhere} \end{cases}$$

The function f_X, f_Y and $f_{X,Y}$ are true density functions; for example

$$f_X(u) \geq 0$$

and

$$\int\limits_{-\infty}^{\infty} f_X(u)du = \int\limits_{0}^{\infty} \frac{du}{(u+1)^2} = -\frac{1}{u+1} \Big|_{0}^{\infty} = 1$$

The joint cumulative distribution function $F_{X,Y}$ is given by

$$F_{X,Y}(x,y) = \int\limits_{-\infty}^{x} \int\limits_{-\infty}^{y} f_X(u) f_Y(v) \, dv \, du$$

or

$$F_{X,Y}(x,y) = \begin{cases} \dfrac{xy}{(1+x)(1+y)} , & x > 0 \text{ and } y > 0 \\ \\ 0 , & \text{elsewhere} \end{cases}$$

The marginal distributions are

$$F_X(x) = F_{X,Y}(x,\infty) = \begin{cases} 0 & , \ x \leq 0 \\ \\ \dfrac{x}{1+x} & , \ x > 0 \end{cases}$$

$$F_Y(y) = F_{X,Y}(\infty,y) = \begin{cases} 0 & , \ y \leq 0 \\ \\ \dfrac{y}{1+y} & , \ y > 0 \end{cases}$$

As expected, the joint distribution function factors into the product of the univariate distribution functions.

3.12 *Conditional Distributions* - The concept of conditional probability was first introduced in Section 1.7. Given the two events A and B, two conditional probabilities were defined by

$$P(A/B) = \frac{P(AB)}{P(B)} \ , \ P(B) \neq 0 \tag{3.12-1}$$

and

$$P(B/A) = \frac{P(AB)}{P(A)} \ , \ P(A) \neq 0 \tag{3.12-2}$$

In the same way, given two random variables X and Y, two conditional distribution functions can be defined by

$$F_{X/Y}(x/y) = \frac{F_{X,Y}(x,y)}{F_Y(y)} \ , \ F_Y(y) \neq 0 \tag{3.12-3}$$

and

$$F_{Y/X}(y/x) = \frac{F_{X,Y}(x,y)}{F_X(x)} \ , \ F_X(x) \neq 0 \tag{3.12-4}$$

where $F_{X,Y}$, F_X, and F_Y are distribution functions that have been defined previously. The function $F_{X/Y}$ is called the *conditional distribution function* of X given $Y=y$; the function $F_{Y/X}$ is called the *conditional distribution function* of Y given $X=x$. It is clear that, in the case where X and Y are independent, we have

$$F_{X/Y}(x/y) = F_X(x) \tag{3.12-5}$$

$$F_{Y/X}(y/x) = F_Y(y) \tag{3.12-6}$$

in agreement with the ideas of Section 3.11.

It is easy to show that $F_{X/Y}$ and $F_{Y/X}$ are univariate distribution functions by showing that they possess the three necessary and sufficient properties discussed in Section 3.2. This proof is left to the reader as an exercise (Problem 12).

It follows from Eqs. (3.12-3) and (3.12-4) that two conditional densities $f_{X/Y}$ and $f_{Y/X}$ may also be defined. In the discrete case, Eq. (3.12-3) may be rearranged to yield

$$F_{X/Y}(x/y)F_Y(y) = \sum_{y \le y_j} f_Y(y) \sum_{x \le x_i} f_{X/Y}(x/y_j)$$

$$= \sum_{y \le y_j} \sum_{x \le x_i} f_{X,Y}(x,y) \qquad (3.12\text{-}7)$$

In other words $f_{X,Y}$ is defined by

$$f_{X/Y}(x/y_j) = \frac{f_{X,Y}(x,y)}{f_Y(y_j)} \quad , \quad f_Y(y_j) \ne 0 \qquad (3.12\text{-}8)$$

that is, $f_Y(y_j) = P\{Y=y_j\}$ is the set of jumps in the distribution function $F_Y(y)$. In the same way

$$f_{Y/X}(y/x_i) = \frac{f_{X,Y}(x,y)}{f_X(x_i)} \quad , \quad f_X(x_i) \ne 0 \qquad (3.12\text{-}9)$$

The function $f_{X/Y}(x/y_j)$ is called the conditional density of X given $Y=y_j$, etc.

In the continuous case, Eq. (3.12-3) may be rewritten as

$$\int_{-\infty}^{y} f_Y(y)\,dy \int_{-\infty}^{x} f_{X/Y}(x/y)\,dx =$$

$$\qquad (3.12\text{-}10)$$

$$\int_{-\infty}^{y} \int_{-\infty}^{x} f_{X,Y}(x,y)\,dx\,dy$$

Differential areas and lengths may be considered to yield

$$P\{y<Y\le y+dy\}P\{x<X\le x+dx \mid y<Y\le y+dy\}$$
$$= P\{x<X\le x+dx \text{ and } y<Y\le y+dy\} \qquad (3.12\text{-}11)$$

or

$$f_Y(y)\,dy\,f_{X/Y}(x/y)\,dx = f_{X,Y}(x,y)\,dx\,dy \qquad (3.12\text{-}12)$$

From this last expression, the conditional density of X given $Y=y$ is

$$f_{X/Y}(x/y) = \frac{f_{X,Y}(x,y)}{f_Y(y)} \quad , \quad f_Y(y) \ne 0 \qquad (3.12\text{-}13)$$

In the same way

$$f_{Y/X}(y/x) = \frac{f_{X,Y}(x,y)}{f_X(x)} \ , \ f_X(x) \neq 0 \qquad (3.12\text{-}14)$$

is the conditional density of Y given $X = x$.

The conditional densities just defined are all proper univariate densities. Consider $f_{X/Y}$ in the continuous case as an example, we have

$$f_{X/Y}(x/y) = \frac{f_{X,Y}(x,y)}{f_Y(y)} \geq 0, \text{ all } x \text{ and } y \qquad (3.12\text{-}15)$$

and

$$\int_{-\infty}^{\infty} f_{X/Y}(x/y) \, dx = \int_{-\infty}^{\infty} \frac{f_{X,Y}(x,y)}{f_Y(y)} \, dx$$

$$= \frac{1}{f_Y(y)} \int_{-\infty}^{\infty} f_{X,Y}(x,y) \, dx = 1 \ (3.12\text{-}16)$$

Example 3.14

Consider the bivariate normal distribution of Example 3.12. This is an example where the random variables X and Y are not independent unless $\rho = 0$. The joint density $\phi_{X,Y}$ and the univariate densities ϕ_X and ϕ_Y are given in Example 3.12. Thus the conditional densities are easily found to be

$$f_{X/Y}(x/y) = \frac{f_{X,Y}(x,y)}{f_Y(y)} = \frac{K \ e^{-Q(x,y)}}{\frac{1}{\sqrt{2\pi}} \ e^{-y^2/2}}$$

or

$$f_{X/Y}(x/y) = \frac{1}{\sqrt{2\pi(1-\rho^2)}} \ e^{-\frac{(x-\rho y)^2}{2(1-\rho^2)}}$$

and

$$f_{Y/X}(y/x) = \frac{1}{\sqrt{2\pi(1-\rho^2)}} \ e^{-\frac{(y-\rho x)^2}{2(1-\rho^2)}}$$

PROBLEMS

1. Plot the c.d.f. and density function for the random variable (a) the sum of the outcomes of two dice, (b) the product of the outcomes of two dice.

2.   Which of the following are discrete density functions:

$$(a)\ p(x) = \begin{cases} 1/12 & \text{for } x = 0 \\ 7/12 & \text{for } x = 1 \\ 1/4 & \text{for } x = 3 \\ 0 & \text{otherwise} \end{cases}$$

$$(b)\ p(x) = \begin{cases} 2^{-|x|} & \text{for } x = \ldots -1,0,1,2,\ldots \\ 0 & \text{otherwise} \end{cases}$$

$$(c)\ p(x) = \begin{cases} (2/3)(1/3)^x & \text{for } x = 0,1,2,\ldots \\ 0 & \text{otherwise} \end{cases}$$

3.   Describe the c.d.f. for the random variable which is the number of times a fair coin is tossed until heads appears. Sketch the c.d.f. for $-4 < x < 4$.

4.   Find the value of A which makes f a density function.

$$f(x) = \begin{cases} Ax & \text{for } 0 \le x \le 4 \\ A(8-x) & \text{for } 4 \le x \le 8 \\ 0 & \text{otherwise} \end{cases}$$

Plot the c.d.f. and density function. What is the probability that $X \le 6$?

5.   Which of the following are density functions?

$$(a)\ f(x) = \begin{cases} \dfrac{1}{b-a} & \text{for } a < x < b \\ 0 & \text{otherwise} \end{cases}$$

$$(b)\ f(x) = \begin{cases} \dfrac{1}{\pi} e^{-|x|} & \text{for } -3 < x < 1 \\ 0 & \text{otherwise} \end{cases}$$

$$(c)\ f(x) = \begin{cases} |x| & \text{for } -1 < x < 1 \\ 0 & \text{otherwise} \end{cases}$$

6. The *triangular distribution* has a density function $f_X(x)$ which is zero except in the interval (a,b). In that interval, it has the form of an isosceles triangle.

   (a) Find an expression for $f_X(x)$ and sketch.

   (b) Find the distribution function $F_X(x)$ and sketch.

7. The *hypergeometric distribution* is defined by Eq. (1.4-11) where $P\{X=k\} = q(k)$ for $k=0,1,...,r$. Prove that

$$f_X(x) = \begin{cases} q(k) & , \ k=0,1,...,r \\ 0 & , \ \text{elsewhere} \end{cases}$$

is a discrete density. Under what restrictions is the binomial density function a good approximation to this density?

8. Construct a distribution function which has an infinite number of jumps in the finite interval (a,b).

9. By plotting on the same graph, see how well the Poisson distribution approximates the binomial distribution for $n=10, p=0.1$, and $\lambda=1.0$; for $n=20, p=0.5$, and $\lambda=1.0$. How should λ be related to n and p?

10. If F is an absolutely continuous distribution function and G is an arbitrary distribution function, show that the convolution denoted by $F*G$ is absolutely continuous. If the density function associated with F is f, show that the density of $F*G$ is

$$d(x) = \int_{-\infty}^{\infty} f(x-y)\, dG(y)$$

11. A distribution function $F_{X,Y}$ is given by

$$F_{X,Y}(x,y) = \int_{-\infty}^{y} \int_{-\infty}^{x} f(u,v)\, du\ dv$$

where the bivariate density function $f(u,v)$ is given by

$$f(u,v) = \begin{cases} 1 & \text{if } 0 \leq v \leq 2u, \ 0 \leq u \leq 1 \\ 0 & \text{elsewhere} \end{cases}$$

Prove that $f(u,v)$ is a density and find the marginal distributions F_X and F_Y.

12. Show that $F_{X/Y}$ and $F_{Y/X}$ as defined by Eqs. (3.12-3) and (3.12-4) are univariate distribution functions.

13. Let X and Y be two independent random variables with distribution functions $F_X(x)$ and $F_Y(x)$ such that

$$F_X(x) \le F_Y(x) \quad , \quad \text{all } x$$

Prove that $P\{X \ge Y\} \ge 1/2$.

14. The *Bernoulli density function* with parameter p is given by

$$f_X(x) = \begin{cases} q=1-p & , & x=0 \\ p & , & x=1 \\ 0 & , & \text{elsewhere} \end{cases}$$

Show that this is indeed a discrete density function and find the corresponding (cumulative) distribution function. Sketch both.

15. The *geometric density function* with parameter p where $0< p <1$ is given by

$$f_X(x) = \begin{cases} p(1-p)^{x-1} & , & x=1,2,\dots \\ 0 & , & \text{otherwise} \end{cases}$$

Show that this is a discrete density function and find the corresponding (cumulative) distribution function. Sketch both. Find an example of a random experiment obeying this law.

16. The *Pascal density function* with parameters r and p is given by

$$f_X(x;r) = \begin{cases} C_x^{x+r-1} \, p^r (1-p)^x & , & x=0,1,2,\dots \\ 0 & , & \text{otherwise} \end{cases}$$

Here $0< p <1$, r is a positive integer, and C_x^{x+r-1} is the binomial coefficient of Eq. (1.4-6). Show that this is a discrete density function and find the corresponding (cumulative) distribution function. Sketch both for several values of the parameter r.

17. Show how the *geometric* density of Problem 15 is related to the Pascal density of Problem 16.

Chapter 4

EXPECTATIONS AND CHARACTERISTIC FUNCTIONS

4.1 *Expectation* - We have been concerned thus far with the development of a model to allow the computation and manipulation of the probabilities of sets of events (outcomes) arising in random experiments. We now seek ways for determining the "average" behavior of these outcomes. A simple example may serve to illustrate what is meant by the term "average".

Example 4.1

You are offered the opportunity to play a game with the following payoff on each outcome: The probability of winning one dollar is $p_1 = 0.3$, of winning three dollars is $p_2 = 0.1$, and of losing one dollar is $p_3 = 0.6$. A reasonable question to ask is whether or not it is profitable to play the game. In other words, what will be the average winnings or *expectation* per game. Assume that the game is played n times and you

> win one dollar $\quad m_1$ times
> win three dollars $\quad m_2$ times
> lose one dollar $\quad m_3$ times

where $n = m_1 + m_2 + m_3$. In n games, the total winnings will be $m_1 + 3m_2 - m_3$; the average winnings per game or *expectation E* is

$$E = \frac{m_1}{n} + 3\frac{m_2}{n} - \frac{m_3}{n}$$

If you accept the relative frequency approach to probability, then this last expression can be approximated by

$$E = p_1 + 3p_2 - p_3$$

If a random variable X is defined as the dollars won on each game, then E can be written as

$$E = \sum_{i=1}^{3} x_i\ P(X = x_i) = \sum_{i=1}^{3} x_i\ p_i$$

———————

Motivated by such considerations as those illustrated in this example, we now proceed to define the expectation operator.

Let X be a random variable and let h be a real-valued function of X. The *expectation* or *expected value* of the function h is given by

$$E[h(X)] = \int_{-\infty}^{\infty} h(x)\, dF_X(x) \qquad (4.1\text{-}1)$$

when the integral exists. The function $F_X(x)$ is the distribution function of X. When X is a continuous random variable, Eq. (4.1-1) can be written as

$$E[h(X)] = \int_{-\infty}^{\infty} h(x)\, f_X(x)\, dx \qquad (4.1\text{-}2)$$

where $f_X(x)$ is the (continuous) density function of X. When X is a discrete random variable, Eq. (4.1-1) becomes

$$E[h(X)] = \sum_i h(x_i)\, f_X(x_i) = \sum_i h(x_i)\, P(X = x_i) \qquad (4.1\text{-}3)$$

where $f_X(x)$ is the (discrete) density function of X.

The expectation has a number of rather obvious and simple properties that follow from the corresponding properties of integrals

(1) $E(h_1(X) + h_2(X)) = E[h_1(X)] + E[h_2(X)] \qquad (4.1\text{-}4)$

(2) $E[c\, h(X)] = c\, E[h(X)] \qquad (4.1\text{-}5)$

These two are *linearity* properties and are a consequence of the fact that expectation (and integration, in general) is a linear operator.

(3) $|E[h(X)]| \le E[|h(X)|] \qquad (4.1\text{-}6)$

(4) $E[h_1(X)] \le E[h_2(X)]$ if $0 \le h_1(X) \le h_2(X) \qquad (4.1\text{-}7)$

(4a) $E[h(X)] \ge 0$ if $h(X) \ge 0 \qquad (4.1\text{-}8)$

(5) $\min[h(X)] \le E[h(X)] \le \max[h(X)] \qquad (4.1\text{-}9)$

These properties are obtained in an elementary way from the definition of expectation. Their proof is left as an exercise (Problem 1).

The linearity property may be extended to a sequence of random variables. Let $X_1, X_2, ..., X_n, ...$ be a sequence such that $E(X_n)$ exists for each X_n. Then the expectation of their sum exists and is the sum of their expectations provided this last sum converges absolutely, that is,

$$E\left(\sum_n X_n\right) = \sum_n E(X_n) \tag{4.1-10}$$

if the right side converges absolutely.

Suppose that X and Y are independent random variables. The expectation of their product can be written as

$$E(XY) = \int_{-\infty}^{\infty} \int_{-\infty}^{\infty} x\ y\ dF_{X,Y}(x,y)$$

$$\tag{4.1-11}$$

$$= \int_{-\infty}^{\infty} \int_{-\infty}^{\infty} xy\ dF_X(x)\ dF_Y(y) = E(X)E(Y)$$

Thus, in the independent case, the expectation of a product is the product of the expectations. More generally, if $X_1, X_2,...,X_n$ are mutually independent random variables and if $h_1, h_2,...,h_n$ are real-valued functions such that

$$E[h_i(X_i)] < \infty\ ,\quad i=1,2,...,n \tag{4.1-12}$$

then

$$E[h_1(X_1)h_2(X_2)...h_n(X_n)]$$

$$\tag{4.1-13}$$

$$= E[h_1(X_1)]E[h_2(X_2)]...E[h_n(X_n)]$$

In the case where a function h depends on more than one random variable it is convenient to adopt vector notation and to define the vector $\mathbf{X}_n$ by

$$\mathbf{X}_n = (X_1, X_2,...,X_n) \tag{4.1-14}$$

If h is a function of $X_1, X_2,...,X_n$, then the expectation of h becomes

$$E[h(\mathbf{X}_n)] = \int_{\mathbf{x}} h(\mathbf{x})\ dF_{\mathbf{X}_n}(\mathbf{x}) \tag{4.1-15}$$

where $\mathbf{x}$ is the vector $\mathbf{x} = (x_1, x_2,...,x_n)$ and where $F_{\mathbf{X}_n}$ is the multivariate distribution function defined in Section 3.10.

The expectation operator has the same properties in the multivariate case as were developed in the beginning of this section for the univariate case. For simplicity consider only bivariate distributions involving the random variables X and Y. We have:

(1) $E[h_1(X,Y)+h_2(X,Y)] = E[h_1(X,Y)]+E[h_2(X,Y)] \tag{4.1-16}$

(2) $E[c\ h(X,Y)] = c\ E[h(X,Y)] \tag{4.1-17}$

(3) $\quad |E[h(X,Y)]| \leq E[|h(X,Y)|]$ $\hfill$ (4.1-18)

(4) $\quad E[h_1(X,Y)] \leq E[h_2(X,Y)]$ if $0 \leq h_1(X,Y) \leq h_2(X,Y)$ (4.1-19)

(4a) $\quad E[h(X,Y) \geq 0$ if $h(X,Y) \geq 0$ $\hfill$ (4.1-20)

(5) $\quad min[h(X,Y)] \leq E[h(X,Y)] \leq max[h(X,Y)]$ $\hfill$ (4.1-21)

These properties may be obtained easily from the corresponding properties of integrals. In addition, if X and Y are *independent*, then, as pointed out in Eq. (4.1-13), there is a sixth property:

(6) $\quad E[h_1(X)h_2(Y)] = E[h_1(X)]E[h_2(Y)]$ $\hfill$ (4.1-22)

4.2 *Moments* - The expectations of powers of random variables are called *moments* and will turn out to be of great usefulness in characterizing distributions. The r-*th* moment (or r-*th* moment about the origin) will be denoted by m_r and is defined by

$$m_r = E(X^r) = \int_{-\infty}^{\infty} x^r \, dF_X(x) \qquad (4.2-1)$$

For the continuous case, this reduces to

$$m_r = E(X^r) = \int_{-\infty}^{\infty} x^r \, f_X(x) \, dx \qquad (4.2-2)$$

and, in the discrete case, to

$$m_r = E(X^r) = \sum_i x_i^r \, f_X(x_i) = \sum_i x_i^r \, P\{X = x_i\} \qquad (4.2-3)$$

The first moment ($r=1$) is called the *mean* and will usually be written as m where

$$m = m_1 = E(X) = \int_{-\infty}^{\infty} x \, dF_X(x) \qquad (4.2-4)$$

In the continuous case, this is just the centroid of the area under the curve $f_X(x)$.

Example 4.2 - Means of Discrete Distributions

The *binomial distribution* of Example 3.3 has a density function $b(x;n,p)$ given by

$$b(x;n,p) = \left(\begin{array}{c} n \\ x \end{array} \right) p^x (1-p)^{n-x} \quad , \quad x = 0,1,...,n$$

for $0 < p < 1$. The mean m of this distribution is

$$m = E(X) = \sum_{x=0}^{n} x \left(\begin{array}{c} n \\ x \end{array} \right) p^{x} (1-p)^{n-x}$$

$$= 0 + np \sum_{x=1}^{n} \left(\begin{array}{c} n-1 \\ x-1 \end{array} \right) p^{x-1} (1-p)^{(n-1)-(x-1)}$$

on changing the index of summation by placing $y = x-1$, we have

$$m = np \sum_{y=0}^{n-1} \left(\begin{array}{c} n-1 \\ y \end{array} \right) p^{y} (1-p)^{n-1-y} = np (p+1-p)^{n-1}$$

$$m = np$$

Thus the mean of the binomial distribution is np.

The *Poisson distribution* of Example 3.4 has a density function $p(x,\lambda)$ given by

$$p(x,\lambda) = \frac{\lambda^{x}}{x!} e^{-\lambda} \quad , \quad x = 0,1,...,n,...$$

for $\lambda > 0$. The mean is

$$m = E(X) = \sum_{x=0}^{\infty} x \frac{\lambda^{x}}{x!} e^{-\lambda} = 0 + \lambda e^{-\lambda} \sum_{x=1}^{\infty} \frac{\lambda^{x-1}}{(x-1)!}$$

$$= \lambda e^{-\lambda} \sum_{y=0}^{\infty} \frac{\lambda^{y}}{y!} = \lambda e^{-\lambda} e^{\lambda} = \lambda$$

Example 4.3 - Means of Continuous Distributions:

The *uniform distribution* of Example 2.5 and Example 3.6 has a density function given by

$$f_X(x) = \begin{cases} \dfrac{1}{b-a} \ , & a \leq x \leq b \\ \\ 0 & \text{elsewhere} \end{cases}$$

The mean of this distribution is

$$m = E(X) = \int_{-\infty}^{\infty} x \, f_X(x) \, dx = \frac{1}{b-a} \int_{a}^{b} x \, dx$$

$$m = \frac{1}{b-a} \frac{x^{2}}{2} \Big|_{a}^{b} = \frac{b^{2}-a^{2}}{2(b-a)} = \frac{b+a}{2}$$

as expected.

The *negative exponential distribution* of Example 3.7 has a density function given by

$$f_X(x) = \begin{cases} e^x & , \quad x \leq 0 \\ 0 & , \quad x > 0 \end{cases}$$

The mean of this distribution is

$$m = E(X) = \int_{-\infty}^{0} x \, e^x \, dx = x \, e^x \Big|_{-\infty}^{0} - \int_{-\infty}^{0} e^x \, dx$$

$$m = -e^x \Big|_{-\infty}^{0} = -1$$

The *unit normal distribution* of Example 3.9 has a density function given by

$$f_x(x) = \frac{1}{\sqrt{2\pi}} \, e^{-x^2/2} \quad , \quad -\infty < x < \infty$$

The mean of this distribution is

$$m = E(X) = \frac{1}{\sqrt{2\pi}} \int_{-\infty}^{\infty} x \, e^{-x^2/2} \, dx = 0$$

since the integrand is odd.

It will be convenient to define a set of moments centered on the mean rather than the origin. The r-*th* moment about the mean (or the r-*th central moment*) will be denoted by u_r and is defined by

$$u_r = E[(x-m)^r] = \int_{-\infty}^{\infty} (x-m)^r \, dF_X(x) \tag{4.2-5}$$

In particular, the *second central moment* is called the *variance,* is frequently written as σ^2, and is given by

$$u_2 = \sigma^2 = E[(X-m)^2] = \int_{-\infty}^{\infty} (x-m)^2 dF_X(x) \tag{4.2-6}$$

The positive square root σ of the variance is called the *standard deviation.*

Both the mean and variance given important information about the distribution of the random variable. The mean is a measure of the "center of gravity" of the distribution. The variance is a measure of the "spread" of the distribution about the mean. If the variance is small, then the distribution is concentrated about the mean. The variance corresponds to the moment of inertia of a mass about its center of gravity.

There are a number of simple properties of moments which will be of considerable use later. These are listed below:

(1) The first moment about the mean is zero; $u_1 = 0$
Proof: $u_1 = E(X-m) = E(X)-E(m) = m - m = 0$ •

(2) The second moment about the origin is equal to the variance plus the square of the mean ; $m_2 = \sigma^2 + m^2$
Proof: $\sigma^2 = E[(X-m)^2] = E(X^2)-2mE(X) + m^2 = m_2 - m^2$ •

(3) The least second moment is the variance σ^2
Proof: We find the value of b for which $E[(X-b)^2]$ is a minimum. Expanding, we have

$$E[(X-b)^2] = E(X^2) - 2bm + b^2 = g(b)$$

A necessary condition of $g(b)$ to be a minimum is

$$\frac{\partial}{\partial b} g(b) = 0 = -2m + 2b,$$

$$b = m$$

The second derivative is positive; therefore $g(b)$ has a minimum at $b = m$:

$$\frac{\partial^2}{\partial b^2} g(b) = 2 \quad •$$

(4) There is a general relationship between moments about the mean (central moments) and moments about the origin
Proof: From the binomial expansion, we obtain

$$(X-m)^r = \sum_{s=0}^{r} (-1)^s \begin{pmatrix} r \\ s \end{pmatrix} X^{r-s} m^s$$

The r-*th* moment about the origin becomes

$$E[(X-m)^r] = \sum_{s=0}^{r} (-1)^s \begin{pmatrix} r \\ s \end{pmatrix} m^s E(X^{r-s})$$

or, finally,

$$u_r = \sum_{s=0}^{r} (-1)^r \begin{pmatrix} r \\ s \end{pmatrix} m^s m_{r-s} \qquad (4.2\text{-}7)$$

For the case where $r = 2$, this expression reduces to Property (2).

One of the most commonly encountered transformations of the random variable X is the *linear transformation* given by

$$Y = aX + b \qquad (4.2\text{-}8)$$

where a and b are constants. In this case the means and variances of X and Y are related in a particularly simple fashion.

Let m_X be the mean of X and let m_Y be the mean of Y. We have

$$E(Y) = m_Y = E(aX + b) = aE(X) + b$$

or

$$m_Y = a\, m_X + b \qquad (4.2\text{-}9)$$

In the same way, the variance of Y is found to be

$$\sigma_Y^2 = E[(Y - u_Y)^2] = E[(aX + b - am_X - b)^2]$$
$$= a^2 E[(X - m_X)^2]$$

or

$$\sigma_Y^2 = a^2\, \sigma_X^2 \qquad (4.2\text{-}10)$$

It is clear from this last result that any random variable Y with finite mean m_Y and finite variance σ_Y^2 may be normalized to have zero mean and unit variance. Define a new random variable X by the linear transformation

$$X = \frac{Y - m_Y}{\sigma_Y} \qquad (4.2\text{-}11)$$

It follows from Eq. (4.2-9) that the mean of X is given by

$$m_X = E(X) = \frac{E(Y) - m_Y}{\sigma_Y} = 0 \qquad (4.2\text{-}12)$$

Also the variance of X is obtained from Eq. (4.2-10) as

$$\sigma_Y^2 = E(X^2) = \frac{\sigma_Y^2}{\sigma_Y^2} = 1 \qquad (4.2\text{-}13)$$

It will often be convenient to work with this normalized random variable $(Y - m_Y)/\sigma_Y$.

It was pointed out in Section 4.1 that for $E[h(x)]$ to exist, the integral or sum defining this expectation must converge absolutely. The following is an example of a continuous distribution whose first moment does not exist.

Example 4.4 - The Cauchy Distribution

The Cauchy distribution is given in Example 3.8 and has a density function

$$f_X(x) = \frac{1}{\pi(1+x^2)} \quad , \quad -\infty < x < \infty$$

The mean of this distribution does not exists since the integral

$$\int_{-\infty}^{\infty} \frac{x}{\pi(1+x^2)} \, dx$$

does not exist, even though the distribution is "centered" on zero.

We now calculate the variances of some of the distributions most commonly encountered.

Example 4.5 - Variances of Discrete Distributions:

Refer to Example 4.2 where the binomial and Poisson distributions are given. For the *binomial distribution,* the variance is found as follows:

$$E\left[X(X-1)\right] = \sum_{x=0}^{n} x(x-1) \left(\begin{array}{c} n \\ x \end{array}\right) p^x (1-p)^{n-x}$$

$$= 0 + 0 + n(n-1)p^2 \sum_{x=2}^{n} \left(\begin{array}{c} n-2 \\ x-2 \end{array}\right) p^{x-2} (1-p)^{n-2-(x-2)}$$

$$= n(n-1)p^2 \sum_{y=0}^{n-2} \left(\begin{array}{c} n-2 \\ y \end{array}\right) p^y (1-p)^{n-2-y}$$

$$= n(n-1)p^2(p+1-p)^{n-2} = n(n-1)p^2$$

We expand $E\left[X(X-1)\right] = n(n-1)p^2$ to obtain

$$E(X^2)-E(X) = n(n-1)p^2$$

It was shown in Example 4.2 that $E(X) = np$; and it follows from Eq. (4.2-7) that

$$\sigma^2 = E(X^2)-[E(X)]^2 = n(n-1)p^2 + np - n^2p^2$$

or

$$\sigma^2 = np(1-p)$$

The variance of the *Poisson distribution* is found in exactly the same way.

$$E\left[X(X-1)\right] = \sum_{x=0}^{\infty} x(x-1) \frac{\lambda^x}{x!} e^{-\lambda}$$

$$= e^{-\lambda} \lambda^2 \sum_{x=2}^{\infty} \frac{\lambda^{x-2}}{(x-2)!} = e^{-\lambda} \lambda^2 e^{\lambda}$$

or

$$E(X^2) - E(X) = \lambda^2$$

Since $E(X) = \lambda$, it follows that

$$\sigma^2 = E(X^2) - [E(X)]^2 = \lambda^2 + \lambda - \lambda^2 = \lambda$$

Thus the mean and variance of the Poisson distribution are both equal to λ.

Example 4.6 - Variances of Continuous Distributions:

We consider the distributions of Example 4.3. For the *uniform distribution,*

$$E(X^2) = \frac{1}{b-a} \int_a^b x^2 \, dx = \frac{b^3 - a^3}{3(b-a)} = \frac{b^2 + ab + a^2}{3}$$

The mean $E(X)$ has already been found to be $(b+a)/2$; hence, the variance is

$$\sigma^2 = E(X^2) - [E(X)]^2 = \frac{(b-a)^2}{12}$$

For the *exponential distribution,*

$$E(X^2) = \int_{-\infty}^0 x^2 e^x \, dx = x^2 e^x \Big|_{-\infty}^0 -2 \int_{-\infty}^0 xe^x \, dx$$

This last integral was evaluated in Example 4.3 and is equal to -1; thus

$$E(X^2) = 2 = \sigma^2 + [E(X)]^2 = \sigma^2 + 1$$
$$\sigma^2 = 1$$

For the *unit normal distribution,* the mean is zero; hence $E(X^2) = \sigma^2$ and

$$\sigma^2 = \frac{1}{\sqrt{2\pi}} \int_{-\infty}^{\infty} x^2 e^{-x^2/2} \, dx$$

On integrating by parts, we obtain

$$\sigma^2 = \frac{-xe^{-x^2/2}}{\sqrt{2\pi}} \Big|_{-\infty}^{\infty} + \frac{1}{\sqrt{2\pi}} \int_{-\infty}^{\infty} e^{-x^2/2} \, dx$$

The last integral has already been shown to be unity in Example 3.9; hence

$$\sigma^2 = 1$$

The unit normal distribution has zero mean and unity variance.

4.3 *The Bienayme-Chebychev Theorem* - We now discuss the first of a number of limit theorems that will prove immensely useful.

Theorem (Bienayme-Chebychev):

Let X be a random variable with $E \mid X \mid^r < \infty$ for $r > 0$ (r is not necessarily an integer). The, for every $\lambda > 0$, it follows that

$$P(\mid X \mid \geq \lambda) \leq \frac{E \mid X \mid^r}{\lambda^r} \qquad (4.3\text{-}1)$$

Proof: The distribution function of X is $F(x) = P(X \leq x)$. We may write

$$E \mid X \mid^r = \int_{-\infty}^{\infty} \mid x \mid^r dF(x)$$

$$= \int_{\mid x \mid < \lambda} \mid x \mid^r dF(x) + \int_{\mid x \mid \geq \lambda} \qquad (4.3\text{-}2)$$

$$\mid x \mid^r dF(x) \geq \int_{x \geq} \lambda \mid x \mid^r dF(x)$$

Since $F(x)$ is nondecreasing, it follows that

$$\int_{x \geq \lambda} \mid x \mid^r dF(x) \geq \lambda^r \int_{x \geq \lambda} dF(x) \qquad (4.3\text{-}3)$$

The last two equations may be combined to yield

$$E \mid X \mid^r \geq \lambda^r \int_{x \geq \lambda} dF(x) = \lambda^r P(\mid X \mid \geq \lambda) \qquad (4.3\text{-}4)$$

and the theorem follows immediately. ●

The special case where $r = 2$ is called *Chebychev's inequality* and may be written in the equivalent forms

$$P(\mid X \mid \geq \lambda) \leq \frac{E \mid X \mid^2}{\lambda^2} \quad , \quad \lambda > 0 \qquad (4.3\text{-}5)$$

or, after a change of variable,

$$P(\mid X - m \mid \geq \lambda \sigma) \leq \frac{1}{\lambda^2} \quad , \quad \lambda > 0 \qquad (4.3\text{-}6)$$

where m is the mean and σ^2 the variance of the random variable X and both are assumed to exist. Equation (4.3-6) shows clearly the close relationship between the variance of a distribution and the dispersion of that distribution about the mean.

Example 4.7

Consider the Chebychev inequality for the case where the mean of a distribution is zero and the variance is unity. In this case

$$P(\mid X \mid \geq \lambda) \leq \frac{1}{\lambda^2}$$

This bound is plotted in Fig. 4.1 and compared to the actual value of $P(\mid X \mid \geq \lambda)$ for the unit normal distribution of Table I at the end of the book.

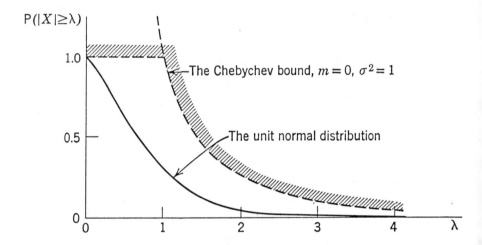

Fig. 4.1 - A comparison of the Chebychev inequality with the unit normal distribution.

As suggested in Fig. 4.1, the Chebychev inequality gives a rather weak bound on most distributions. On the other hand, it applies to *all* distributions whose means and variances exist. The only information required to apply the inequality is the mean and variance. This inequality will turn out to be a very powerful tool in treating other limit theorems, particularly those involving sequences of random variables.

A somewhat more general form of Eq. (4.3-1) is easily obtained.

Theorem (Generalized Bienayme-Chebychev):

Let $g(x)$ be a non-decreasing non-negative function defined on $(0,\infty)$. Then, for $\lambda \geq 0$,

$$P\{\mid X \mid \geq \lambda\} \leq E\{g(\mid X \mid)\}/g(\lambda) \qquad (4.3\text{-}7)$$

whenever the right side exists.

Proof: Let $F(x)$ be the distribution function of the random variable X and form

$$E\{g(|X|)\} \geq \int_{-\infty}^{-\lambda} G(x)\, dF(x) + \int_{\lambda-0}^{\infty} g(x)\, dF(x)$$

$$\geq g(\lambda)\left[\int_{-\infty}^{-\lambda} dF(x) + \int_{\lambda-0}^{\infty} dF(x)\right] \qquad (4.3\text{-}8)$$

or

$$E\{g(|X|)\} \geq g(\lambda) P\{|X| \geq \lambda\} \qquad \bullet$$

The Bienayme-Chebychev theorem is obtained by setting $g(x) = |x|^r$.

4.4 *The Moment Generating Function* - An expectation which is closely related to the moments of a distribution is the *moment generating function $M_X(t)$* defined by

$$M_X(t) = E(e^{tX}) = \int_{-\infty}^{\infty} e^{tx}\, dF(x) \qquad (4.4\text{-}1)$$

where t is a real variable defined on $(-\infty,\infty)$. Since e^{tX} has the power series expansion

$$e^{tx} = 1 + tx + \frac{t^2 x^2}{2!} + \cdots + \frac{t^n x^n}{n!} + \cdots \qquad (4.4\text{-}2)$$

Eq. (4.4-1) may be written as

$$M_X(t) = 1 + tE(X) + \frac{t^2}{2!} E(X^2) \qquad (4.4\text{-}3)$$

$$+ \cdots + \frac{t^n}{n!} E(X^n) + \cdots$$

or

$$M_X(t) = 1 + tm_1 + \frac{t^2}{2!} m_2 + \cdots + \frac{t^n}{n!} m_n + \cdots \qquad (4.4\text{-}4)$$

It is apparent that the moments $m_i = E(X^i)$ are easily found from $M_X(t)$ since

$$M_X(0) = 1 \qquad (4.4\text{-}5)$$

$$\frac{\partial}{\partial t} M_X(t)\Big|_{t=0} = m_1 \qquad (4.4\text{-}6)$$

$$\frac{\partial^n}{\partial t^n} M_X(t)\Big|_{t=0} = m_n \qquad (4.4\text{-}7)$$

Thus $M_X(t)$ generates the moments of X and any particular moment m_n is easily found by differentiation if $M_X(t)$ is known.

Let $M_X(t)$ and $M_Y(t)$ be the moment generating functions of X and Y, respectively. For the case where

$$Y = aX + b \qquad (4.4\text{-}8)$$

it is clear that

$$M_Y(t) = E(e^{tY}) = E(e^{taX} e^{tb}) = e^{bt} E(e^{atX}) \qquad (4.4\text{-}9)$$

or

$$M_Y(t) = e^{bt} M_X(at) \qquad (4.4\text{-}10)$$

Example 4.8 - Moment Generating Functions of Some Discrete Distributions:

We consider again the binomial and Poisson distributions of Example 4.2.

The binomial distribution

$$f_X(x) = b(x;n,p) = \left(\begin{array}{c} n \\ x \end{array} \right) p^x (1-p)^{n-x} \quad , \qquad x = 0,1,\ldots,n$$

$$M_X(t) = E(e^{tX}) = \sum_{x=0}^{n} e^{tx} \left(\begin{array}{c} n \\ x \end{array} \right) p^x (1-p)^{n-x}$$

$$= \sum_{x=0}^{n} \left(\begin{array}{c} n \\ x \end{array} \right) (p\, e^t)^x (1-p)^{n-x}$$

or

$$M_X(t) = (p\, e^t + 1 - p)^n$$

Note that

$$M_X(0) = 1$$

$$\frac{\partial M_X(t)}{\partial t} \bigg|_{t=0} = n(p\, e^t + 1 - p)^{n-1} p e^t \bigg|_{t=0} = np = m$$

$$\frac{\partial^2 M_X(t)}{\partial t^2} \bigg|_{t=0} = np(1-p) + n^2 p^2 = \sigma^2 + m^2$$

etc.

The Poisson distribution

$$f_X(x) = p(x,\lambda) = \frac{\lambda^x}{x!} e^{-\lambda} \quad , \quad x = 0,1,\ldots,n,\ldots$$

$$M_X(t) = E(e^{tX}) = \sum_{x=0}^{\infty} e^{tx} \frac{\lambda^x}{x!} e^{-\lambda}$$

$$= e^{-\lambda} \sum_{x=0}^{\infty} \frac{(\lambda e^t)^x}{x!} = e^{-\lambda} e^{\lambda e^t}$$

or

$$M_X(t) = e^{\lambda(e^t - 1)}$$

Again, we note that

$$M_X(0) = 1$$

$$\frac{\partial M_X(t)}{\partial t} \Big|_{t=0} = e^{\lambda(e^t-1)} \lambda e^t \Big|_{t=0} = \lambda = m$$

$$\frac{\partial^2 M_X(t)}{\partial t^2} \Big|_{t=0} = \lambda + \lambda^2 = E(X^2) = \sigma^2 + m^2$$

etc.

Example 4.9 - The Moment Generating Function for the Unit Normal Distribution:

The density function for the unit normal distribution is given by

$$f_z(x) = \frac{1}{\sqrt{2\pi}} e^{-x^2/2} \quad , \quad -\infty < x < \infty$$

The moment generating function $M_X(t)$ is

$$M_X(t) = E(e^{tX}) = \frac{1}{\sqrt{2\pi}} \int_{-\infty}^{\infty} e^{tx} e^{-x^2/2} dx$$

We complete the square in x in the exponent to obtain

$$M_X(t) = e^{t^2/2} \left[\frac{1}{\sqrt{2\pi}} \int_{-\infty}^{\infty} e^{-(x-t)^2/2} dx \right]$$

The term in brackets has a value of unity as is easily seen if the change in variable $y = x - t$ is made; hence

$$M_X(t) = e^{t^2/2}$$

A joint moment generating function $M_{\mathbf{X}_n}$ can be defined for the random vector $\mathbf{X}_n = (X_1, X_2, ..., X_n)$ as

$$M_{\mathbf{X}_n}(\mathbf{t}_n) = E(e^{(\mathbf{t}_n, \mathbf{X}_n)}) \qquad (4.4\text{-}11)$$

where $\mathbf{t}_n = (t_1, t_2, ..., t_n)$ and

$$(\mathbf{t}_n, \mathbf{X}_n) = \sum_{i=1}^{n} t_i X_i \qquad (4.4\text{-}12)$$

This moment generating function may be written more explicitly as

$$M_{\mathbf{X}_n}(\mathbf{t}_n) = \int_{-\infty}^{\infty} \cdots \int_{-\infty}^{\infty} e^{(\mathbf{t}_n, \mathbf{X}_n)} \, dF_{\mathbf{X}_n}(\mathbf{x}_n) \qquad (4.4\text{-}13)$$

where $\mathbf{x}_n = (x_1, x_2, ..., x_n)$ and $F_{\mathbf{X}_n}$ is the joint distribution function of $\mathbf{X}_n$. In the bivariate case where $X_1 = X$ and $X_2 = Y$, this expression becomes

$$M_{X,Y}(t_1, t_2) = E(e^{(t_1 X + t_2 Y)})$$

$$\qquad (4.4\text{-}14)$$

$$= \int_{-\infty}^{\infty} \int_{-\infty}^{\infty} e^{t_1 x + t_2 y} \, dF_{X,Y}(x, y)$$

Consider the case where X and Y are independent random variables with moment generating functions $M_X(t)$ and $M_Y(t)$. It is desired to find the moment generating function $M_Z(t)$ of the random variable Z where

$$Z = X + Y \qquad (4.4\text{-}15)$$

The moment generating function M_Z is defined by

$$M_Z(t) = E(e^{tZ}) = E(e^{t(X+Y)}) \qquad (4.4\text{-}16)$$

Since X and Y are independent, the expectation of the product is the product of the expectations, or

$$M_Z(t) = E(e^{tX}) E(e^{tY}) = M_X(t) M_Y(t) \qquad (4.4\text{-}17)$$

More generally, let Z_n be the sum

$$Z_n = \sum_{i=1}^{n} X_i \qquad (4.4\text{-}18)$$

where the X_i are mutually independent random variables with moment generating functions $M_{X_i}(t) = E(e^{tX_i})$. Then it is easily seen that the moment generating function of Z_n is given by the product

$$M_{Z_n}(t) = \prod_{i=1}^{n} M_{X_i}(t) \qquad (4.4\text{-}19)$$

4.5 *The Chernoff Bound* - This inequality provides, in general, a tighter bound than the Chebychev inequality. In addition, it applies to sums of independent random variables having the same distribution. Its disadvantage is that it requires a knowledge of the moment generating function of the distribution. Let $X_1, X_2, ..., X_n$ be a sequence of independent and identically distributed (i.i.d.) random variables and form the sum

$$Y_n = \sum_{i=1}^{n} X_i \qquad (4.5\text{-}1)$$

Let $M(t) = E(e^{tX_i})$ be the common moment generating function of the X_i. Let λ be a real number and define the events A and B by

$$A = \{Y_n \geq \lambda\} \qquad (4.5\text{-}2)$$

$$B = \{Y_n \leq \lambda\} \qquad (4.5\text{-}3)$$

Let I_A and I_B be the indicators for A and B, respectively. It is clear that

$$e^{tY_n} \geq e^{t\lambda} I_A \quad , \quad t \geq 0 \qquad (4.5\text{-}4)$$

and that

$$e^{tY_n} \geq e^{t\lambda} I_B \quad , \quad t \leq 0 \qquad (4.5\text{-}5)$$

These last two expressions are readily verified by taking the logarithm of both sides of the inequality. Compare to Eqs. (4.3-2) and (4.3-3).

Since the X_i are independent, each with moment generating function $M(t)$, the moment generating function of Y_n is simply

$$M_{Y_n}(t) = E(e^{tY_n}) = [M(t)]^n \qquad (4.5\text{-}6)$$

On taking the expectations of both sides of Eqs. (4.5-4) and (4.5-5), we obtain

$$[M(t)]^n \geq e^{t\lambda} P(A) \quad , \quad t \geq 0 \qquad (4.5\text{-}7)$$

$$[M(t)]^n \geq e^{t\lambda} P(B) \quad , \quad t \leq 0 \qquad (4.5\text{-}8)$$

For convenience in notation define the function $g(t)$ by

$$g(t) = \log_e M(t) \qquad (4.5\text{-}9)$$

Now Eqs. (4.5-7) and (4.5-8) may be rearranged to yield

$$P(A) = P\{Y_n \geq \lambda\} \leq e^{-t\lambda + ng(t)} \quad , \quad t \geq 0 \qquad (4.5\text{-}10)$$

$$P(B) = P\{Y_n \leq \lambda\} \leq e^{-t\lambda + ng(t)} \quad , \quad t \leq 0 \qquad (4.5\text{-}11)$$

It is now desired to find the value of t which makes the right side of these last two expressions a minimum. We differentiate with respect to t and set the result equal to zero:

$$\frac{\partial}{\partial t}[e^{-t\lambda+ng(t)}] = 0 = e^{-t\lambda+ng(t)}[-\lambda+ng'(t)] \qquad (4.5\text{-}12)$$

Thus t should be a root of the equation

$$g'(t) = \frac{\lambda}{n} \quad \text{or} \quad \lambda = ng'(t) \qquad (4.5\text{-}13)$$

where $g'(t)$ is the derivative of $g(t)$ with respect to t. It is easy to show that Eq. (4.5-13) yields a minimum by showing that the derivative of Eq. (4.5-12) is greater than zero. Denote the root of Eq. (4.5-13) by t_r and substitute this equation into Eqs. (4.5-10) and (4.5-11) to obtain

$$P\{Y_n \geq ng'(t_r)\} \leq e^{-n[t_r g'(t_r)-g(t_r)]} \quad , \quad t_r \geq 0 \qquad (4.5\text{-}14)$$

$$P\{Y_n \leq ng'(t_r)\} \leq e^{-n[t_r g'(t_r)-g(t_r)]} \quad , \quad t_r \leq 0 \qquad (4.5\text{-}15)$$

Recall that $g(t) = \log_e M(t)$. These last two expressions are called the *Chernoff Bound*.

Example 4.10

Let us consider the unit normal distribution with density function

$$\phi_X(x) = \frac{1}{\sqrt{2\pi}} e^{-x^2/2} \quad , \quad -\infty < x < \infty$$

In this case $n=1$ so that $Y_n = X$. It has already been shown in Example 4.9 that the moment generating function is $M(t) = e^{t^2/2}$; thus

$$g(t) = \log_e M(t) = t^2/2$$

and

$$g'(t) = t$$

The single root of Eq. (4.5-13) is

$$t_r = \lambda$$

Now the Chernoff bound for the unit normal distribution may be written as

$$P\{X \geq \lambda\} \leq e^{-\lambda^2/2} \quad , \quad \lambda \geq 0$$

$$P\{X \leq \lambda\} \leq e^{-\lambda^2/2} \quad , \quad \lambda \leq 0$$

It is interesting to compare this bound to the Chebychev inequality of Fig. 4.1 (see Problem 6).

4.6 *The Characteristic Function* - Although most of the discussion thus far concerning the description of random variables has been formulated in terms of distribution functions or density functions, such functions may not always be easy to work with directly. In many cases, results can be obtained more easily in terms of characteristic functions. If X is a random variable, its *characteristic function* $M_X(iu)$ is defined as

$$M_X(iu) = E(e^{iuX}) = \int_{-\infty}^{\infty} e^{iux} \, dF_x(x) \tag{4.6-1}$$

where $i = \sqrt{-1}$, u is a real variable on $(-\infty,\infty)$, and F_X is the distribution function of X. Formally, the characteristic function is obtained from the moment generating function $M_X(t)$ by the substitution $t = iu$.

There are several advantages in working with the characteristic function instead of the moment generating function. In the first place, $M_X(t)$ may not exist since the integral of Eq. (4.4-1) may not converge absolutely. On the other hand, the absolute convergence of Eq. (4.6-1) is assured since $|e^{iux}| = 1$. In the second place, it follows from the theory of Fourier transforms that Eq. (4.6-1) may be inverted so that the distribution function may be obtained from the characteristic function. Thus a knowledge of either the characteristic function or the distribution function is equivalent to a knowledge of the other.

The characteristic function possesses most of the same properties as the moment generating function. In many cases, these may be derived by the substitution $t = iu$. Thus,

(1) $M_X(0) = 1$ $\hspace{3cm}$ (4.6-2)

(2) The moments of the random variable X are given by

$$m_n = \frac{\partial^n}{\partial(iu)^n} M_X(iu) \Big|_{u=0} = (-1)^n \frac{\partial^n}{\partial u^n} M_X(iu) \Big|_{u=0} \tag{4.6-3}$$

(3) For the linear change in variable $Y = aX + b$, we have

$$M_Y(iu) = e^{iub} M_X(aiu) \tag{4.6-4}$$

Additional properties not possessed by the moment generating function include:

(4) $|M_X(iu)| \leq 1$ for all u $\hspace{3cm}$ (4.6-5)

Proof:

$$|M_X(iu)| \leq \int_{-\infty}^{\infty} |e^{iux}| \, dF_X(x) = 1 \quad \bullet$$

(5)$M_X(-iu) = M*_X(iu)$ (4.6-6)

where $M*_X$ indicates the complex conjugate.

Proof:

$$M_X(-iu) = \int_{-\infty}^{\infty} e^{-iux} \, dF_X(x) = M*_X(iu) \quad \bullet$$

(6) Every characteristic function is uniformly continuous** on the real line $-\infty < u < \infty$.

Proof: For $\epsilon > 0$, define a number b by

$$1 - F_X(b) < \epsilon/4$$

$$F_X(-b) < \epsilon/4$$

Form the difference

$$| M_X[i(u+\delta)] - M_X(iu) | \; = \; | \int_{-\infty}^{\infty} (e^{ix(u+\delta)} - e^{iux}) \, dF_X(x) |$$

$$= \; | \int_{-\infty}^{\infty} e^{iux} (e^{i\delta x} - 1) \, dF_X(x) |$$

$$\leq \int_{-\infty}^{\infty} | e^{i\delta x} - 1 | \, dF_X(x)$$

From the identity $| e^{i\delta x} - 1 | = 2 | \sin\dfrac{\delta x}{2} | \leq 2$, we have

$$| M_X[i(u+\delta)] - M_X(iu) | \; \leq 2 \int_{-\infty}^{-b} dF_X(x) + 2 \int_{b}^{\infty} dF_X(x)$$

$$+ \; 2 \int_{-b}^{b} | \sin\frac{\delta x}{2} | \, dF_X(x)$$

The last integral is independent of u and can be made arbitrarily small by making δ arbitrarily small; thus

$$\lim_{\delta \to 0} | M_X[i(u+\delta)] - M_X(iu) | \leq \epsilon$$

uniformly in u and so $M_X(iu)$ is uniformly continuous on $(-\infty, \infty)$ $\bullet$.

**A function $F(t)$ is *uniformly continuous* on $[a,b]$ if, given $\epsilon > 0$, there exists a $\delta > 0$ such that whenever $t_1, t_2 \in [a,b]$ and $| t_1 - t_2 | < \delta$, then $| f(t_1) - f(t_2) | < \epsilon$. A function which is uniformly continuous on $[a,b]$ is obviously continuous at each point of $[a,b]$, but the converse is not true. The idea in uniform continuity is that $f(t_1)$ can be made close to $f(t_2)$ by making t_1 close to t_2 simultaneously for all $t_1, t_2 \in [a,b]$. The choice of δ does not depend on t_1 and t_2 but only on their difference. See Section 8.1 for a discussion of the analogous idea of uniform convergence.

Let us consider the case where X is a continuous random variable with density function $f_X(x)$. In this case Eq. (4.6-1) becomes

$$M_x(iu) = \int_{-\infty}^{\infty} e^{iux} f_X(x) \, dx \qquad (4.6\text{-}7)$$

Since the right side of this equation converges absolutely, the theory of Fourier transforms insures that it can be inverted to yield:

$$f_X(x) = \frac{1}{2\pi} \int_{-\infty}^{\infty} M_x(iu) e^{-iux} \, du \qquad (4.6\text{-}8)$$

These last two equations form a Fourier transform pair and extensive tables exist [Campbell and Foster (1948)] giving corresponding pairs. In most treatments of the application of Fourier transforms, Eq. (4.6-8) would be called the *direct transform* and Eq. (4.6-7) the *inverse transform*. Also the factor $1/2\pi$ is usually associated with Eq. (4.6-7) rather than with Eq. (4.6-8).

If X is a discrete random variable taking on the sequence of values $x_1, x_2, \ldots$, then Eq. (4.6-1) becomes the series

$$\phi_X(iu) = \sum_i e^{iux_i} P(X=x_i) = \sum_i e^{iux_i} f_X(x_i) \qquad (4.6\text{-}9)$$

where f_X is the discrete density function of x. In this case the inversion relationship becomes

$$\lim_{M \to \infty} \frac{1}{2M} \int_{-M}^{M} \phi_X(iu) e^{-iux} \, du = \begin{cases} f_X(x_i), & x=x_i, i=1,2,\ldots \\ 0, & x \neq x_i \end{cases} \qquad (4.6\text{-}10)$$

The great advantage of the Fourier transform relationship between the density function and the characteristic function lies in the fact that the uniqueness of Fourier transforms insures that working with one of these functions is equivalent to working with the other. In many applications the difficulties encountered in analysis may be much less with one function than with the other.

Example 4.11 - The Characteristic Functions of Some Common Distributions:

These characteristic functions are derived in exactly the same way as were the moment generating functions of Examples 4.8 and 4.9. We list the Fourier pairs below:

The Binomial Distribution

$$b(x;n,p) = \binom{n}{x} p^x (1-p)^{n-x}, \quad x=0,1,\ldots,n$$

$$M_X(iu) = (p \, e^{iu} + 1-p)^n$$

The Poisson Distribution

$$f_X(x) = p(x,\lambda) = \frac{\lambda^x}{x!} e^{-\lambda} \quad, \quad x = 0,1,...,n,...$$

$$M_X(iu) = e^{\lambda(e^{iu}-1)}$$

The Unit Normal Distribution

$$f_X(x) = \frac{1}{\sqrt{2\pi}} e^{-x^2/2} \quad, \quad -\infty < x < \infty$$

$$M_X(iu) = e^{-u^2/2}$$

The joint characteristic function for the random vector $\mathbf{X}_n = (X_1,X_2,...,X_n)$ is defined as

$$M_{\mathbf{X}_n}(i\,\mathbf{u}_n) = E(e^{i(\mathbf{u}_n,\mathbf{X}_n)}) \tag{4.6-11}$$

where $\mathbf{u}_n = (u_1,u_2,...,u_n)$ and, as before

$$(\mathbf{u}_n,\mathbf{X}_n) = \sum_{i=1}^{n} u_i\, X_i \tag{4.6-12}$$

It is clear that this joint characteristic function has a set of properties equivalent to those derived in the univariate case; that is,

(1) $M_{\mathbf{X}_n}(\mathbf{O}) = 0 \quad$, where $\mathbf{O} = (0,0,...,0)$ $\tag{4.6-13}$

(2) $|M_{\mathbf{X}_n}(i\,\mathbf{u}_n)| \leq 1$ for all $\mathbf{u}_n \in R_n$ $\tag{4.6-14}$

As with the moment generating function, the joint characteristic function of independent random variables has a particularly simple form. If X and Y are independent random variables with characteristic functions $M_X(iu)$ and $M_Y(iu)$, then, as in Eq. (4.4-17), the characteristic function $M_X(iu)$ of $Z = X+Y$ is

$$M_Z(iu) = E\{e^{iu(X+Y)}\} = E\{e^{iuX}\}E\{e^{iuY}\}$$
$$= M_X(iu)M_Y(iu) \tag{4.6-15}$$

In the same way, for the sum Z_n of mutually independent random variables X_j given by Eq. (4.4-18), the joint characteristic function $M_{Z_n}(iu)$ is

$$M_{Z_n}(iu) = \prod_{j=1}^{n} M_{X_j}(iu) \tag{4.6-16}$$

4.7 *Covariances and Correlation Coefficients* - Those moments, such as means and variances, that have been considered thus far relate only to single random variables. Consider the case where X and Y are two random variables that are not necessarily independent. Let the means of these random variables be $E(X) = m_X$ and $E(Y) = m_Y$ and the variances be $E[(X-m_X)^2] = \sigma_X^2$ and $E[(Y-m_Y)^2] = \sigma_Y^2$. As in the univariate case, the rs- *th moment about the origin* may be defined as

$$m_{rs} = E(X^r Y^s) = \int_{-\infty}^{\infty} \int_{-\infty}^{\infty} x^r y^s \, dF_{X,Y}(x,y) \qquad (4.7\text{-}1)$$

In particular the means of X and Y are

$$m_{10} = m_X = E(X) \qquad (4.7\text{-}2)$$

and

$$m_{01} = m_Y = E(Y) \qquad (4.7\text{-}3)$$

In the same way, the rs- *th central moment* is

$$u_{rs} = E[(X-m_X)^r (Y-m_Y)^s] \qquad (4.7\text{-}4)$$

In particular the variance of X is

$$u_{20} = \sigma_X^2 = E[(X-m_X)^2] \qquad (4.7\text{-}5)$$

and the variance of Y is

$$u_{02} = \sigma_Y^2 = E[(Y-m_Y)^2] \qquad (4.7\text{-}6)$$

In addition to these two variances, there is a third related possibility, the second mixed moment u_{11} which is called the *covariance:*

$$u_{11} = cov(X,Y) = \sigma_{X,Y} = E[(X-m_X)(Y-m_Y)] \qquad (4.7\text{-}7)$$

We note in passing that the covariance is zero if X and Y are independent since, in that case,

$$\sigma_{X,Y} = E(X-m_X)E(Y-m_Y) = 0 \qquad (4.7\text{-}8)$$

It should be emphasized, however, that the converse is not true, that is, *uncorrelated* random variables (random variables where $\sigma_{X,Y} = 0$) are *not necessarily independent.*

It is often convenient to normalize the covariance, thus defining the *correlation coefficient* by

$$\rho_{X,Y} = \frac{\sigma_{X,Y}}{\sigma_X \sigma_Y} \qquad (4.7\text{-}9)$$

where σ_X and σ_Y are the positive square roots of the variances of X and Y. It is easy to show that $\rho_{X,Y}$ is bounded in magnitude by unity, that is,

$$-1 \leq \rho_{X,Y} \leq 1 \qquad (4.7\text{-}10)$$

Let us consider the non-negative function

$$E\{[(X-m_X)a + (Y-m_Y)b]^2\} \geq 0 \qquad (4.7\text{-}11)$$

On expanding and taking the expectation term by term, we obtain

$$\sigma_X^2 a^2 + 2\sigma_{X,Y}\ ab + \sigma_Y^2 b^2 \geq 0 \qquad (4.7\text{-}12)$$

The left side of this expression is a homogeneous quadratic form in a and b and could be written as

$$Aa^2 + B\,2ab + Cb^2 \geq 0 \qquad (4.7\text{-}13)$$

where A and C are non-negative. The condition that this expression be non-negative is that it have no real roots; that is, that

$$AC - B^2 = \sigma_X^2 \sigma_Y^2 - \sigma_{X,Y}^2 \geq 0 \qquad (4.7\text{-}14)$$

or that

$$\rho_{X,Y}^2 = \frac{\sigma_{X,Y}^2}{\sigma_X^2 \sigma_Y^2} \leq 1 \qquad (4.7\text{-}15)$$

This last expression is equivalent to Eq. (4.7-10).

If X and Y are linearly related, then the equality holds in Eq. (4.7-10). Let

$$X = bY \qquad (4.7\text{-}16)$$

where b is an arbitrary real non-zero number. It is clear that

$$\rho_{X,Y} = \frac{E[(bY-bm_Y)(Y-m_Y)]}{|b|\ \sigma_Y\sigma_Y} = \frac{b}{|b|} \qquad (4.7\text{-}17)$$

Thus $\rho_{X,Y}$ is given by

$$\rho_{X,Y} = \begin{cases} +1\ ,\ b > 0 \\ -1\ ,\ b < 0 \end{cases} \qquad (4.7\text{-}18)$$

when $X = bY$.

Let us consider now the mean and variance of a linear combination of random variables. let the random variable Y be defined by

$$Y = \sum_i a_i\ X_i \qquad (4.7\text{-}19)$$

where the X_i are arbitrary random variables with means m_i and variances σ_i^2. The mean of Y is given by

$$m_Y = E(Y) = \sum_i a_i\ E(X_i) = \sum_i a_i\ m_i \qquad (4.7\text{-}20)$$

as has been previously established. The variance σ_Y^2 becomes

$$\sigma_Y^2 = E[(Y-m_Y)^2] = E\{[\sum_i a_i(X_i-m_i)]^2\} \qquad (4.7\text{-}21)$$

or

$$\sigma_Y^2 = \sum_i a_i^2 \sigma_i^2 + \sum_i \sum_j a_i \, a_j \, \sigma_{ij} \qquad (4.7\text{-}22)$$
$$\scriptstyle i \ne j$$

where $\sigma_{ij} = E\left[(X_i - m_i)(X_j - m_j)\right]$ is the covariance of X_i and X_j when $i \ne j$. If the X_i are *uncorrelated;* that is, if

$$\sigma_{ij} = 0 \quad \text{all } i,j \text{ except } i = j \qquad (4.7\text{-}23)$$

then Eq. (4.7-22) reduces to

$$\sigma_Y^2 = \sum_i a_i^2 \sigma_i^2 \qquad (4.7\text{-}24)$$

A *sufficient* condition for the last two equations to hold is that the X_i be mutually independent. On the other hand, as mentioned earlier, random variables may be uncorrelated without being independent. Sometimes a set of random variables that obey Eq. (4.7-23) is called *linearly independent.*

For the univariate case, the mean m, the variance σ^2, and the second moment about the origin m_2 are related by

$$m_2 = \sigma^2 + m^2 \qquad (4.7\text{-}25)$$

as proved in Section 4.2. In the bivariate case, an analogous relationship can be derived by expanding Eq. (4.7-7) to obtain

$$\sigma_{X,Y} = u_{11} = E(XY) - m_X \, m_Y \qquad (4.7\text{-}26)$$

or

$$E(XY) = \sigma_{X,Y} + m_X \, m_Y \qquad (4.7\text{-}27)$$

If $X = Y$, this expression reduces to Eq. (4.7-25).

It was pointed out in Section 4.2 that the least second moment is the variance. A more general problem is the following. For what values of a and b is

$$E\left[(Y - aX - b)^2\right] = \text{a minimum?} \qquad (4.7\text{-}28)$$

Solving Eq. (4.7-28) is equivalent to finding the values of a and b such that $aX + b$ best approximates Y in the sense that the mean-squared error [given by Eq. (4.7-28)] is a minimum. Let the function A be defined by $A(a,b) = E\left[(Y - aX - b)^2\right]$. A set of *necessary* conditions for this mean-squared error A to be a minimum with respect to a and b is

$$\frac{\partial}{\partial a} A(a,b) = 0 = -E(XY) + a \, E(X^2) + b \, E(X) \qquad (4.7\text{-}29)$$

and

$$\frac{\partial}{\partial b} A(a,b) = 0 = -E(Y) + a \, E(X) + b \qquad (4.7\text{-}30)$$

These two equations may be solved simultaneously for a and b to yield

$$a = \frac{E(XY)-E(X)E(Y)}{E(X^2)-[E(X)]^2} = \frac{\sigma_{X,Y}}{\sigma_X^2} = \frac{\sigma_Y}{\sigma_X}\rho_{X,Y} \qquad (4.7\text{-}31)$$

and

$$b = E(Y) - aE(X) \qquad (4.7\text{-}32)$$

These last two equations for a and b do in fact minimize the mean-squared error $A(a,b)$ of Eq. (4.7-28). To show this, we consider $A(\alpha,\beta)$ where α and β are any two real numbers. It follows from Eq. (4.7-28) that

$$\begin{aligned} A(\alpha,\beta) &= E\{[Y-(\alpha X+\beta)]^2\} \\ &= E\{[Y-(aX+b)+(aX+b)-\alpha X+\beta)]^2\} \\ &= E\{[Y-aX+b)]^2\} \qquad (4.7\text{-}33) \\ &\quad + 2E\{[Y-(aX+b)][(aX+b)-(\alpha X+\beta)]\} \\ &\quad + E\{[(a-\alpha)X+(b-\beta)]^2\} \end{aligned}$$

This expression may be rearranged to yield

$$\begin{aligned} A(\alpha,\beta) &= A(a,b)+2(a-\alpha)E\{[Y-(aX+b)]X\} \\ &\quad + 2(b-\beta)E[Y-(aX+b)] \qquad (4.7\text{-}34) \\ &\quad + E\{[(a-\alpha)X+(b-\beta)]^2\} \end{aligned}$$

If Eqs. (4.7-31) and (4.7-32) are satisfied, then the two middle terms in this last expression are zero and $A(\alpha,\beta)$ may be written as

$$A(\alpha,\beta) = A(a,b)+E\{[(a-\alpha)X+(b-\beta)]^2\} \qquad (4.7\text{-}35)$$

Note that the last term is non-negative unless $\alpha{=}a$ and $\beta{=}b$, in which case it is zero. Thus

$$A(\alpha,\beta) \geq A(a,b) \qquad (4.7\text{-}36)$$

unless $\alpha{=}a$ and $\beta{=}b$. In other words, Eqs. (4.7-31) and (4.7-32) yield a minimum mean-squared error for Eq. (4.7-28).

We have thus solved the problem of finding the least mean-squared error linear approximation of Y by X. The line $y=ax+b$ is called a *regression line*.

4.8 *Conditional Expectation* - The notion of conditional expectation is a very important concept in least-mean-squared-error prediction and in the theory of random processes. In fact, it is one of the most fundamental tools in probability theory.

Let X be a random variable with finite expectation $E(X)$ defined on the probability field $(\Omega,\mathbf{B},P)$. Let A be an event such that $P(A){>}0$. It has already been pointed out in Section 3.12 that the *conditional distribution function* of the random variable X under the condition A is

$$F_{X/A}(x/A) = P(X \leq x/A) \tag{4.8-1}$$

In Section 3.12, for two random variables X and Y, the conditional distributions $F_{X/Y}$ and $F_{Y/X}$ and the conditional densities $f_{X/Y}$ and $f_{Y/X}$ were defined. In terms of these definitions, the *conditional expectation* or *conditional mean* of Y given $X=x$ is defined by

$$E(Y/x) = \int_{-\infty}^{\infty} y \; dF_{Y/X}(y/x) \tag{4.8-2}$$

The quantity $E(Y/x)$ is a function of x. In the same way, the conditional expectation of X given $Y=y$ is the function of y defined by

$$E(X/y) = \int_{-\infty}^{\infty} x \; dF_{X/Y}(x/y) \tag{4.8-3}$$

More generally, let $h(x,y)$ be a real-valued function whose domain is the real plane. The conditional expectation of $h(X,Y)$ given that $X=x$ is

$$E[h(X,Y)/x] = \int_{-\infty}^{\infty} h(x,y) \; dF_{Y/X}(y/x) \tag{4.8-4}$$

In the case where X and Y are continuous random variables, this last expression becomes

$$E[h(X,Y)/x] = \int_{-\infty}^{\infty} h(x,y) \; f_{Y/X}(y/x) dy \tag{4.8-5}$$

When X and Y are discrete, we have

$$E[h(X,Y)/x_i] = \sum_j h(x_i,y_j) f_{Y/X}(y_j/x_i) \tag{4.8-6}$$

where the $\{x_i\}$ are the set of jumps in F_X and the $\{y_j\}$ are the set of jumps in F_Y. Similar definitions exist for $E[h(X,Y)/y]$.

It is clear that an expression such as Eq. (4.8-2) is a function of x. On the other hand, if x varies randomly according to the probability law governing the random variable X, then the expression $E(Y/X)$ is properly treated as a random variable. In this case it is appropriate to consider the expectation of this random variable and write

$$E[E(Y/X)] = \int_{-\infty}^{\infty} [\int_{-\infty}^{\infty} y \; dF_{Y/X}(y/x)] \; dF_X(x) \tag{4.8-7}$$

This expression may be rearranged with the aid of Eq. (3.12-4) to yield

$$E\left[E\left(Y/X\right)\right] = \int\limits_{-\infty}^{\infty} \int\limits_{-\infty}^{\infty} y \; dF_{X,Y}\left(x,y\right)$$

$$\text{(4.8-8)}$$

$$= \int\limits_{-\infty}^{\infty} y \; dF_Y\left(y\right) = E\left(Y\right)$$

Thus, the expected value of the conditional expected value is the unconditional expected value. It is easy to see that, in the more general case

$$E\left\{E\left[h\left(X,Y\right)/X\right]\right\} = \int\limits_{-\infty}^{\infty} \int\limits_{-\infty}^{\infty} h\left(x,y\right) \; dF_{X,Y}\left(x,y\right)$$

$$= E\left[h\left(X,Y\right)\right] \qquad \text{(4.8-9)}$$

The random variable $E\left(Y/X\right)$ will be of considerable interest later. Its mean is just $E\left(Y\right)$ as given by Eq. (4.8-8). Its variance will be denoted by $VarE\left(Y/X\right)$ and is given by

$$Var \; E\left(Y/X\right) = E\left\{\left[E\left(Y/X\right)-E\left(Y\right)\right]^2\right\} \qquad \text{(4.8-10)}$$

Let us consider now the variance of Y and write

$$Var \; Y = \sigma_Y^2 = E\left\{\left[Y-E\left(Y\right)\right]^2\right\}$$

$$\text{(4.8-11)}$$

$$= E\left[E\left\{\left[Y-E\left(Y\right)\right]^2/X\right\}\right]$$

The inner expectation may be written as

$$E\left\{\left[Y-E\left(Y/X\right) + E\left(Y/X\right)-E\left(Y\right)\right]^2/X\right\} \qquad \text{(4.8-12)}$$

Let

$$A = Y-E\left(Y/X\right)$$

and

$$B = E\left(Y/X\right)-E\left(Y\right).$$

The conditional expectation given by Eq. (4.8-12) may be evaluated term-by-term to yield

$$E\left(A^2/X\right) = E\left\{\left[Y-E\left(Y/X\right)\right]^2/X\right\} = \sigma_{Y/X}^2 \qquad \text{(4.8-13)}$$

$$E\left(B^2/X\right) = B^2/X \qquad \text{(4.8-14)}$$

$$E\left(2AB/X\right) = 2\left(B/X\right)E\left\{\left[Y-E\left(Y/X\right)\right]/X\right\} = 0 \qquad \text{(4.8-15)}$$

On taking the outer expectation in Eq. (4.8-11) we obtain a decomposition of σ_Y^2 as

$$\sigma_Y^2 = E\left(\sigma_{Y/X}^2\right) + E\left\{[E\left(Y/X\right)-E\left(Y\right)]^2/X\right\} \qquad (4.8\text{-}16)$$

It follows from Eq. (4.8-10) that this last expression can be written as

$$\sigma_Y^2 = E\left(\sigma_{Y/X}^2\right) + \mathrm{Var}\ E\left(Y/X\right) \qquad (4.8\text{-}17)$$

We shall return to this expression when we discuss least mean-squared error prediction in the next section.

Example 4.12

The bivariate normal distribution in normalized form has been encountered in Examples 3.12 and 3.14. The conditional density function $\phi_{Y/X}$ was derived in Example 3.14. The conditional mean of Y is just

$$E\left(Y/x\right) = \frac{1}{\sqrt{2\pi(1-\rho^2)}} \int_{-\infty}^{\infty} y\ e^{-\left[\frac{(y-\rho x)^2}{2(1-\rho^2)}\right]}\ dy$$

On making the change of variable

$$z = \frac{y-\rho x}{\sqrt{1-\rho^2}} \quad , \quad dy = \sqrt{1-\rho^2}\ dz$$

we obtain

$$E\left(Y/x\right) = \frac{\sqrt{1-\rho^2}}{\sqrt{2\pi}} \int_{-\infty}^{\infty} z\ e^{-z^2/2}\ dz + \rho x\ \frac{1}{\sqrt{2\pi}} \int_{-\infty}^{\infty} e^{-z^2/2}\ dz$$

The first integral is zero since the integrand is odd; the second integral is $\sqrt{2\pi}$; thus

$$E\left(Y/x\right) = \rho x$$

and, as expected,

$$E\left[E\left(Y/X\right)\right] = E\left(\rho X\right) = \rho E\left(X\right) = 0 = E\left(Y\right)$$

The conditional variance $\mathrm{Var}\ Y/x = \sigma_{Y/x}^2$ is

$$\sigma_{Y/x}^2 = E\left[(Y-\rho x)^2/x\right]$$

$$= \frac{1}{\sqrt{2\pi(1-\rho^2)}} \int_{-\infty}^{\infty} (y-\rho x)^2\ e^{-\left[\frac{(y-\rho x)^2}{2(1-\rho^2)}\right]}\ dy$$

Changing the variable to z as before, we have

$$\sigma_{Y/x}^2 = (1-\rho^2)\ \frac{1}{\sqrt{2\pi}} \int_{-\infty}^{\infty} z^2\ e^{-z^2/2}\ dz = 1-\rho^2$$

from the properties of the unit normal distribution. Thus

$$\sigma_{Y/x}^2 = 1-\rho^2 \le 1 = \sigma_Y^2$$

where $\sigma_Y^2 = 1$ is the variance of the unit normal distribution.

This concept of conditional expectation can be extended in a straightforward fashion to the case where more than two random variables are involved. As in previous cases, let $\mathbf{X}_n$ be the random vector

$$\mathbf{X}_n = (X_1, X_2, ..., X_n) \qquad (4.8\text{-}18)$$

Let $\mathbf{Y}_m = (X_{i_1}, X_{i_2}, ..., X_{i_m})$ be a vector whose components are an arbitrary subset of the X_i with $m < n$. Let $h(\mathbf{Y}_m)$ be a real-valued function whose domain is R_m. Let $\mathbf{Z}_p = (X_{j_1}, X_{j_2}, ..., X_{j_p})$ be a vector whose components are any subset of the remaining n-m variables X_i. We define the conditional expectation of $h(\mathbf{Y}_m)$ given $X_{j1} = x_{j1}, X_{j2} = x_{j2}, ..., X_{jp} = x_{jp}$ by

$$E[h(\mathbf{Y}_m)/\mathbf{Z}_p] = \int_{-\infty}^{\infty} \cdots \int_{-\infty}^{\infty} h(\mathbf{y}_m)dF_{\mathbf{Y}_m/\mathbf{Z}_p}(\mathbf{y}_m/\mathbf{z}_p) \qquad (4.8\text{-}19)$$

where $\mathbf{y}_m$ and $\mathbf{z}_p$ are defined in an obvious manner. As in the bivariate case, it is clear that

$$E\{E[h(\mathbf{Y}_m)/\mathbf{Z}_p]\} = E[h(\mathbf{Y}_m)] \qquad (4.8\text{-}20)$$

4.9 *Least Mean-Squared Error Prediction* - In Section 4.7, a *linear* least mean-squared prediction problem was solved. The linear estimate $\tilde{Y} = aX + b$ of the random variable Y was found such that the mean-squared error $E[(Y-\tilde{Y})^2]$ was a minimum. The values of the constants a and b were given by Eqs. (4.7-31) and (4.7-32) and involved the five parameters $E(X)$, $E(Y)$, σ_X^2, σ_Y^2, and $\rho_{X,Y}$. Note that the equation $\tilde{Y} = aX + b$ can be arranged in the symmetrical form

$$\frac{\tilde{Y}-E(Y)}{\sigma_Y} = \rho_{X,Y} \frac{X-E(X)}{\sigma_X} \qquad (4.9\text{-}1)$$

In more general situations, it may be desired to approximate Y by some function $\tilde{Y} = N(X)$ which is not necessarily linear such that the mean-squared error

$$L = E\{[Y-N(X)]^2\} \qquad (4.9\text{-}2)$$

is a minimum. The mean-squared error L can be rewritten as

$$L = E[E\{[Y-N(X)]^2/X\}] \qquad (4.9\text{-}3)$$

Let us now rewrite the inner expectation as in Eq. (4.8-12) to obtain

$$E\{[Y-E(Y/X) + E(Y/X)-N(X)]^2/X\} \qquad (4.9\text{-}4)$$

where A has been previously defined as
$$A = Y - E(Y/x)$$
and C is defined by

$$C = E(Y/X) - N(X).$$

As before

$$E(A^2/X) = \sigma_{Y/X}^2 \qquad (4.9\text{-}5)$$

$$E(C^2/X) = C^2/X \qquad (4.9\text{-}6)$$

$$E(2AC/X) = 2(C/X)E\{[E(Y) - E(Y/X)]/X\} = 0 \qquad (4.9\text{-}7)$$

On taking the outer expectation in Eq. (4.9-3), we obtain

$$L = E(\sigma_{Y/X}^2) + E\{[E(Y/X) - N(X)]^2/X\} \qquad (4.9\text{-}8)$$

Since the first term in this expression is non-negative, it is clear that L is a minimum when the second term is zero; that is, when

$$N(X) = E(Y/X) = \tilde{Y} \qquad (4.9\text{-}9)$$

The function $E(Y/X)$ is called the *regression function* of Y on X and is the least mean-squared error estimate of Y in terms of X.

It was pointed out that the *linear* regression formula of Eq. (4.9-1) required five parameters. It is apparent that the general expression of Eq. (4.9-9) requires a great many more, since the conditional density function $f_{Y/X}(y/x)$ must be known for every x. For some $f_{Y/X}$, Eq. (4.9-9) may be a linear function of X; in this case, the least mean-squared error predictor is linear. A *sufficient* condition for this to be true is that X and Y be jointly normal. Suppose, for example, that X and Y have the joint normal distribution of Example 3.12. Then $E(Y/X)$ is given in Example 4.12 as

$$\tilde{Y} = E(Y/X) = \rho X \qquad (4.9\text{-}10)$$

and the least mean-squared error predictor is linear. It is not *necessary*, however, that X and Y be jointly normal for $\tilde{Y}$ to be a linear function of X.

Let us consider now the error which results with the least mean-squared error predictor. This predictor is given by Eq. (4.9-9) and, when this equation is satisfied, the mean-squared error $L_{\min}$ is given by the first term of Eq. (4.9-8):

$$L_{\min} = E(\sigma_{Y/X}^2) \qquad (4.9\text{-}11)$$

Now Eq. (4.8-17) may be used to rewrite this last expression as

$$L_{\min} = \sigma_Y^2 - Var\ E(Y/X) \qquad (4.9\text{-}12)$$

Note that both terms on the right side of this equation are non-negative. Also $L_{\min}$ is non-negative since it is a mean-squared error. As a consequence we have

$$L_{\min} \geq 0 \qquad (4.9\text{-}13)$$

$$\max L_{\min} = \sigma_Y^2 \qquad (4.9\text{-}14)$$

$$L_{min} = 0 \quad \text{iff.} \quad \sigma_{\hat{Y}}^2 = Var\ E(Y/X) \qquad (4.9\text{-}15)$$

Let us consider the error in the linear case where $\tilde{Y}$ is constrained to be of the form $\tilde{Y} = aX + b$ with a and b given by Eqs. (4.7-31) and (4.7-32). Keep in mind that $aX + b$ may not be equal to $E(Y/X)$; hence $\tilde{Y}$ may not be the least mean-squared error predictor but only the least *linear* mean-squared error predictor. Suppose $a = 0$ so that we find the best linear predictor $\tilde{Y}_1$ of Y without recourse to X. Then

$$\tilde{Y}_1 = b = E(Y) \qquad (4.9\text{-}16)$$

The error L_1 is just

$$L_1 = \sigma_Y^2 \qquad (4.9\text{-}17)$$

and is the maximum error according to Eq. (4.9-14)

In the general *linear* case, the mean-squared error is easily found by expanding Eq. (4.7-28) with the values of a and b substituted from Eqs. (4.7-31) and (4.7-32). We obtain

$$L = E\{[Y - E(Y)] - \frac{\sigma_Y}{\sigma_X}\rho_{X,Y}[X - E(X)]\}^2 = E\{[G - H]^2\}$$

$$= E(G^2) - 2E(GH) + E(H^2) \qquad (4.9\text{-}18)$$

where

$$E(G^2) = \sigma_Y^2 \qquad (4.9\text{-}19)$$

$$-2E(GH) = -2\frac{\sigma_Y}{\sigma_X}\rho_{X,Y}\ E\{[Y - E(Y)][X - E(X)]\}$$

$$= -2\sigma_Y^2\ \rho_{X,Y}^2 \qquad (4.9\text{-}20)$$

$$E(H^2) = \frac{\sigma_Y^2}{\sigma_X^2}\rho_{X,Y}^2\ E\{[X - E(X)]^2\} \qquad (4.9\text{-}21)$$

$$= \sigma_Y^2 \rho_{X,Y}^2$$

Thus the mean-squared error in the linear case is the simple expression

$$L = (1 - \rho_{X,Y}^2)\ \sigma_Y^2 \qquad (4.9\text{-}22)$$

when $\rho_{X,Y} = 0$ the error is a maximum since there is no contribution from X. When $\rho_{X,Y} = \pm 1$, X and Y are linearly related (as shown previously) and the linear predictor is perfect; this results in a least mean-squared error of zero.

PROBLEMS

1. Prove the validity of Eqs. (4.1-4) through (4.1-9).
2. If $E\,|X| < \infty$, show that it is given by

$$E\,|X| = \int_{-\infty}^{0} F(x)\,dx + \int_{0}^{\infty} [1-F(x)]\,dx$$

   where F is the distribution function of X.
3. Suppose that X is a continuous random variable with distribution function F. Find $E\,[F(X)]$.
4. The *median* α of a continuous random variable X is defined by

$$\frac{1}{2} = \int_{-\infty}^{\alpha} f_X(x)\,dx = \int_{\alpha}^{\infty} f_X(x)\,dx$$

   Show that $E\,|X-b|$ is a minimum where $b = \alpha$.
5. Prove Markov's inequality which is stated as follows: If X is a *non-negative* random variable, then

$$P(X \geq \lambda) \leq \frac{E(X)}{\lambda}$$

   for all $\lambda > 0$.
6. For the unit normal distribution, compare the Chernoff and Chebychev bounds by plotting them on the same graph as in Fig. 4.1.
7. Find the moment generating function and characteristic function for a random variable which is uniformly distributed in the interval (a,b).
8. Find the moment generating function and characteristic function for a random variable with the negative exponential density

$$f_X(x) = \begin{cases} e^x & , \ x \leq 0 \\ 0 & , \ x > 0 \end{cases}$$

   Calculate the first four moments using the moment generating function; repeat by direct calculation.
9. Let $X_1, X_2, ..., X_n$ be a sequence of independent random variables each with the unit normal density function

$$f_X(x) = \frac{1}{\sqrt{2\pi}}\,e^{-x^2/2} \quad , \quad -\infty < x < \infty$$

   Find the density function and the mean and variance of the sum $Y = \sum_{i=1}^{n} X_i$. (Hint: use characteristic functions).

10. Let X be a continuous random variable whose density function $f_X(x)$ vanishes for negative x. Show that the mean of X is given by

$$E(X) = \int_0^\infty [1-F_X(x)] \, dx$$

11. Let two random variables Y and Z be defined by
$$Y = aX+b$$

$$Z = cX+d$$
where a, b, c, and d are constants. Under what conditions on a, b, c, and d are Y and Z uncorrelated?

12. Construct an example where $\rho_{X,Y} = 0$ but X and Y are not independent.

13. Show that the correlation coefficient $\rho_{X,Y}$ is independent of any linear change of variables.

14. If $M(iu)$ is a characteristic function, show that the following are also characteristic functions:

    a) $[2-M(iu)]^{-1}$

    b) $|M(iu)|^2$

15. Another generating function which is sometimes used is

$$G(t) = E(t^X)$$

Show how $G(t)$ may be used to generate moments. Find $G(t)$ for the Poisson distribution and verify the first two moments.

16. A function $h(x)$ is twice differentiable and has a second derivative which is non-negative everywhere. Show that

$$h[E(X)] \le E[h(X)]$$

for an arbitrary random variable X. Verify for a particular function $h(x)$ and for a random variable with a given distribution.

17. Bill, Fred and Al take turns tossing a coin, in that order, until a head appears. The man who first tosses a head keeps the coin. Find a set of entrance fees to make the game fair; that is, to give equal expected gain for all players.

18. If the random variable X is the product of the two outcomes resulting from the tossing of a pair of dice, what is the mean of X? What is the most probable value of X?

19. Prove that

$$f_X(x) = \frac{1}{\cosh \pi x} \quad , \quad -\infty < x < \infty$$

is a density function. Find the corresponding characteristic function.

20. The Chebychev inequality of Eq. (4.3-1) may be generalized to two dimensions. Let X and Y be random variables with zero means, variances of unity, and correlation coefficient ρ given by

$$\rho = E\{XY\}/\sqrt{E\{X^2\}E\{Y^2\}}$$

Show that

$$E\{max\,[X^2,Y^2]\} \leq 1 + \sqrt{1-\rho^2}$$

21. Derive the moment generating function (or characteristic function) for (a) the Bernoulli density of Problem 3.14, (b) the geometric density of Problem 3.15, and (c) the Pascal density of Problem 3.16. Find the means and variances for each of the three densities.

Chapter 5

THE BINOMIAL, POISSON, AND NORMAL DISTRIBUTIONS

In this chapter we consider in some detail the binomial, Poisson, and normal distributions. Although the binomial distribution has its own applications, part of its importance lies in the fact that both the Poisson and normal distributions can be derived as limiting cases.

5.1 *The Binomial Distribution* - In many of the simplest situations where probability concepts are to be applied, it happens that there is a sequence of repeated independent trials with only two possible outcomes on each trial and with the probabilities of these two outcomes fixed throughout the trials. Such sequences are called *Bernoulli trials*. Perhaps the commonest example is a sequence of coin-tossings, where a typical sequence might be

HTTHTHH .....

It is conventional to call one of the two outcomes S (success) with probability of occurrence given by p and the other outcome F (failure) with probability $q = 1-p$. Since the individual trials are independent by definition, the probabilities multiply and the probability of any given sequence is the product obtained when S is replaced by p and F by $q = 1-p$. Thus, for the previous example, if H is success, the probability P of the sequence is

$$P = p(1-p)(1-p)p(1-p)pp.....$$ (5.1-1)

or, for a sequence of length n,

$$P = p^x(1-p)^{n-x}$$ (5.1-2)

where x is the number of successes and $(n-x)$ is the number of failures.

Example 5.1

Consider the successive tosses of a true die. Let throwing a two be considered success and all other outcomes failure. Then the probability P of the Bernoulli sequence F2FFFF2, where F stands for any number 1,3,4,5,6 is

$$P = (\frac{1}{6})^2(\frac{5}{6})^5$$

Suppose that we are interested in the probability of a given number of successes (say x successes) in n Bernoulli trials. One way to obtain x successes would be to have the sequence

$$\underbrace{SSS \ldots\ldots S}_{\text{X}} \qquad \underbrace{FFF \ldots\ldots F}_{\text{n - X}}$$

where the first x trials are successes and the last$(n-x)$ are failures. The probability P of this sequence has already been given by Eq (5.1-2). It is apparent that any other sequence containing exactly x successes will be acceptable. In fact there are available n cells or slots into which x successes can be put (The remaining cells each have a failure placed in them in each case.). The number of different ways in which this can be done is just the number of combinations of n things taken x at a time or

$$\begin{pmatrix} n \\ x \end{pmatrix} = \frac{n!}{(n-x)!x!} \qquad (5.1\text{-}3)$$

Each of these $\begin{pmatrix} n \\ x \end{pmatrix}$ sequences has a probability given by Eq. (5.1-2). Consequently the probability of exactly x successes in n Bernoulli trials is

$$b(x;n,p) = \begin{pmatrix} n \\ x \end{pmatrix} p^x(1-p)^{n-x} \quad , \quad x = 0,1,\ldots,n \qquad (5.1\text{-}4)$$

where p is the probability of success in a single trial.

The quantity $b(x;n,p)$ has already been encountered in a number of examples and is called the *binomial density function*. It is the $x-th$ term in the binomial expansion of $(p+q)^n$

$$(p+q)^n = \sum_{x=0}^{n} \begin{pmatrix} n \\ x \end{pmatrix} p^x q^{n-x} = \sum_{x=0}^{n} b(x;n,p) \qquad (5.1\text{-}5)$$

It is apparent that $q = 1-p$ and, consequently, that

$$[p+(1-p)]^n = \sum_{x=0}^{n} b(x;n,p) = 1 \qquad (5.1\text{-}6)$$

Since, in addition, $b(x;n,p)$ is non-negative, it is a discrete density function.

The mean m of the binomial distribution has already been calculated in Example 4.2 and is given by

$$m = E(X) = np \qquad (5.1\text{-}7)$$

This result is intuitively reasonable. In a large number of independent trials with probability p of success on each trial, we would expect approximately np successes. The variance σ^2 of this distribution was calculated in Example 4.5 and was shown to be

$$\sigma^2 = np(1-p) \qquad (5.1\text{-}8)$$

If other moments of the binomial distribution are desired, it will be simpler to deal with the moment generating function. This function $M_X(t)$ was shown in Example 4.8 to be

$$M_X(t) = E(e^{tX}) = (p \ e^t + 1-p)^n \qquad (5.1\text{-}9)$$

For large n, it will often be convenient to work with the normalized random variable

$$Y = X/n \qquad (5.1\text{-}10)$$

which cannot exceed unity. For example, if X is the number of heads observed in n tosses of a coin, then Y is the proportion of heads. The mean of Y is

$$E(Y) = E(X/n) = \frac{1}{n} E(X) = np/n = p \qquad (5.1\text{-}11)$$

and the variance of Y follows from Eq. (4.2-10)

$$\sigma_Y^2 = np(1-p)/n^2 = p(1-p)/n \qquad (5.1\text{-}12)$$

This last expression indicates that, as n becomes large, the distribution of Y clusters more and more about the mean p. A question of some interest is: How close to the mean p is Y for large n? Since Y is a random variable, the question can be answered only in a probabilistic sense. We have, from the Chebychev inequality of (4.3-6),

$$P\{|Y-p| \geq \epsilon\} \leq \sigma_Y^2/\epsilon^2 \qquad (5.1\text{-}13)$$

for any arbitrary $\epsilon > 0$. Since σ_Y^2 is given by Eq. (5.1-12), we write

$$\lim_{n \to \infty} P\{|Y-p| \geq \epsilon\} = \lim_{n \to \infty} \frac{p(1-p)}{n \epsilon^2} = 0 \qquad (5.1\text{-}14)$$

for $\epsilon > 0$. Thus, in a probabilistic sense, the random variable Y can be made to approach the mean arbitrarily closely by making the number of trials n large enough. We say that Y *converges to* p *in probability* when Eq. (5.1-14) is satisfied. The concept of the convergence of random variables will be discussed in some detail in Chapter 8.

The fact that the normalized random variable $Y = X/n$ in a sequence of Bernoulli trials convergence in probability to the mean p is called *Bernoulli's Theorem*. It is a special case of the *Law of Large Numbers* which will be discussed in Chapter 8.

Bernoulli's Theorem suggests that the binomial density function $b(x;n,p)$ has a maximum in the vicinity of its mean np. Let us find the *mode or central term*, that is, the value of x for which $b(x;n,p)$ is a maximum. Consider the ratio

$$\frac{b\,(x\,;n\,,p\,)}{b\,(x-1;n\,,p\,)} = \frac{\left(\begin{array}{c} n \\ x \end{array}\right)p^{\,x}\,(1-p\,)^{\,n-x}}{\left(\begin{array}{c} n \\ x-1 \end{array}\right)p^{\,x-1}(1-p\,)^{\,n-x+1}}$$

$$= \frac{(n-x+1)p}{x\,(1-p\,)} = 1 + \frac{(n+1)p-x}{x\,(1-p\,)} \qquad (5.1\text{-}15)$$

Since $x\,(1-p\,)$ is a non-negative quantity, the ratio in this last equation obeys the inequalities

$$\frac{b\,(x\,;n\,,p\,)}{b\,(x-1;n\,,p\,)} = \left\{ \begin{array}{ll} > 1 \text{ if } x < (n+1)p \\ = 1 \text{ if } x = (n+1)p \\ < 1 \text{ if } x > (n+1)p \end{array} \right. \qquad (5.1\text{-}16)$$

In other words $b\,(x\,;n\,,p\,)$ is greater than the preceding term if $x < (n+1)p$ and is less than the preceding term if $x > (n+1)p$. If $x = (n+1)p$ is an integer, then there are two central terms which are equal. Thus, the density function $b\,(x\,;n\,,p\,)$ should be a maximum when $x = (n+1)p$. If $(n+1)p$ is not an integer, then the integer next below $(n+1)p$ is the value of x for which $b\,(x\,;n\,,p\,)$ is a maximum. Note that this integer is the most probable number of successes in the sequence of n Bernoulli trials, but the corresponding probability $b\,(x\,;n\,,p\,)$ may be very small.

Example 5.2

Let $n = 5$ and $p = 1/2$, so that the binomial density function is given by

$$b\,(x\,;5,1/2) = \frac{1}{32} \left(\begin{array}{c} 5 \\ x \end{array}\right) \quad, \quad x = 0,1,2,3,4,5$$

or

$$b\,(0;5,1/2) = b\,(5;5,1/2) = 1/32$$
$$b\,(1;5,1/2) = b\,(4;5,1/2) = 5/32$$
$$b\,(2;5,1/2) = b\,(3;5,1/2) = 10/32$$

This density function is plotted in Fig. 5.1. It is seen to have two central terms at $x = 2$ and $x = 3$.

Example 5.3

Consider the random experiment which consists of 40 tosses of an unbiased coin. The random variable X is the number of heads obtained in the 40 tosses. The density function governing this experiment is $b\,(x\,;40,1/2)$. Since $(n+1)p = 20.5$, the most probable number of heads is 20. However, the probability of exactly 20 heads in 40 tosses is

$$b\,(20;40,1/2) = \frac{40!}{20!20!}\,(\frac{1}{2})^{40}$$

which is an extremely small number (see Problem 5).

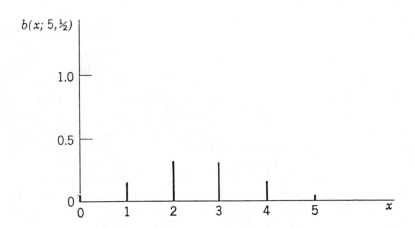

Fig. 5.1 - The binomial density function $b(x;5,1/2)$

5.2 *The Poisson Distribution* - This distribution can be obtained as a limiting approximation to the binomial distribution. Suppose the situation exists where n is very large and p is very small but the product $np = \lambda$ is of reasonable size. More precisely, let n approach infinity and p approach zero in such a way that the product λ exists and is non-zero. For this case an approximation to the binomial distribution can be found in a form which is amenable to calculation and manipulation.

Consider the term $b\,(0;n\,,p\,)$:

$$b\,(0;n\,,p\,) = \left[\begin{array}{c} n \\ 0 \end{array}\right] p^{0}(1-p\,)^{n} = (1-p\,)^{n} \qquad (5.2\text{-}1)$$

Replace p by λ/n so that

$$\lim_{n \to \infty} b\,(0;n\,,p\,) = \lim_{n \to \infty} (1-\frac{\lambda}{n})^{n} = e^{-\lambda} \qquad (5.2\text{-}2)$$

since

$$\lim_{x \to \infty} (1+\frac{1}{x})^{x} = e \qquad (5.2\text{-}3)$$

Thus, for large n, the term $b(0;n,p)$ is given approximately by $e^{-\lambda}$ where $\lambda = np$. In the same way

$$b(1;n,p) = \begin{pmatrix} n \\ 1 \end{pmatrix} p(1-p)^{n-1} = \frac{np}{1-p} b(0;n,p) \qquad (5.2\text{-}4)$$

and

$$\lim_{n \to \infty} b(1;n,p) = \lim_{n \to \infty} \frac{\lambda}{1-\lambda/n} e^{-\lambda} = \lambda e^{-\lambda} \qquad (5.2\text{-}5)$$

We continue and find that

$$\lim_{n \to \infty} b(2;n,p) = \frac{\lambda^2}{2} e^{-\lambda} \qquad (5.2\text{-}6)$$

Finally

$$\lim_{n \to \infty} b(x;n,p) = p(x;\lambda) = \frac{\lambda^x}{x!} e^{-\lambda} \qquad (5.2\text{-}7)$$

if p approaches zero in such a way that $np = \lambda$. For large n, Eq. (5.2-7) with $\lambda = np$ is called the *Poisson approximation to the binomial distribution,* and, in the limit as $n \to \infty$, this is the *Poisson distribution.*

We note that

$$\sum_{x=0}^{\infty} p(x;\lambda) = e^{-\lambda} \sum_{x=0}^{\infty} \frac{\lambda^x}{x!} \qquad (5.2\text{-}8)$$

or, from the power series expansion of e^{λ},

$$\sum_{x=0}^{\infty} p(x;\lambda) = e^{-\lambda} e^{\lambda} = 1 \qquad (5.2\text{-}9)$$

Since $p(x,\lambda)$ is also non-negative, it is a density function.

The mean of the Poisson distribution was found in Example 4.2 and is

$$m = E(X) = \lambda \qquad (5.2\text{-}10)$$

as might be expected. The variance was found in Example 4.5 to be

$$\sigma_X^2 = E[(X-\lambda)^2] = \lambda \qquad (5.2\text{-}11)$$

Thus the variance and the mean for the Poisson distribution are both equal to λ. The moment generating function was calculated in Example 4.8 and is

$$M_X(t) = e^{\lambda(e^t - 1)} \qquad (5.2\text{-}12)$$

5.3 *The Normal or Gaussian Distribution* - Suppose that the random variable X is binomially distributed so that its density function is

$$f_X(x) = b(x;n,p) = \begin{pmatrix} n \\ x \end{pmatrix} p^x (1-p)^{n-x} \ , \ x = 0,1,...,n \ (5.3\text{-}1)$$

In the limit as $n \to \infty$, it can be shown [Cramer (1946)] that the density function $f_X(x)$ approaches

$$f_X(x) = \frac{1}{\sqrt{np(1-p)}\sqrt{2\pi}} \ e^{-\frac{(x-np)^2}{2np(1-p)}} \qquad (5.3\text{-}2)$$

in the sense that the ratio of (5.3.1) and (5.3.2) approaches unity. In the limit, the random variable X is continuous $(-\infty,\infty)$ and*

$$f_X(x) = \frac{1}{\sqrt{2\pi}\sigma} \ e^{-(x-m)^2/2\sigma^2} \ , \ -\infty < x < \infty \qquad (5.3\text{-}3)$$

where

$$m = np$$

$$\sigma^2 = np(1-p)$$

As discussed by Feller (1968), the approximation is amazingly good for small n $(n \geq 5)$, as long as p is not too far from $1/2$.

Equation (5.3-3) is the density function for the *normal or Gaussian distribution* with mean m and variance σ^2. It will sometimes be convenient to refer to this distribution by the notation $N(m,\sigma)$. In the case where $m=0$ and $\sigma^2=1$, we have the *unit normal distribution* $N(0,1)$ with density

$$\phi_X(x) = \frac{1}{\sqrt{2\pi}} \ e^{-x^2/2} \ , \ -\infty < x < \infty \qquad (5.3\text{-}4)$$

This distribution has already been used extensively in the examples, and in Example 4.9 it was shown that $\phi_X(x) \geq 0$ and that

$$\int_{-\infty}^{\infty} \phi_X(x) \ dx = 1 \qquad (5.3\text{-}5)$$

Both the density function and the distribution function were plotted in Fig. 3.5. In addition Table I in the Appendix lists numerical values for these two functions. For comparison, a plot of Eq. (5.3-3) is shown in Fig. 5.2 for a fixed value of the mean m and for three values of the standard deviation σ_1, σ_2 and σ_3.

---

*As we have defined them, both m and σ^2 are infinite in the limit. It is necessary, of course, for Eq. (5.3-3) to be renormalized so that m and σ^2 are finite. See Feller (1968) and Dubes (1968) for detailed discussion.

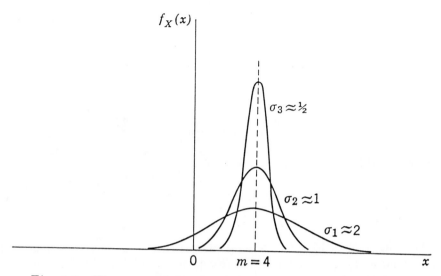

Fig. 5.2 - The normal density function for non-zero mean
and different standard deviations

That Eq. (5.3-3) is indeed a density function is easily shown
by noting that $f_X(x) \geq 0$ and forming the integral

$$\int_{-\infty}^{\infty} f_X(x)\,dx = \frac{1}{\sqrt{2\pi}\,\sigma} \int_{-\infty}^{\infty} e^{-(x-m)^2/2\sigma^2}\,dx \qquad (5.3\text{-}6)$$

Now make the linear change of variable $y = (x-m)/\sigma$ and this
last expression becomes

$$\int_{-\infty}^{\infty} f_X(x)\,dx = \frac{1}{2\pi} \int_{-\infty}^{\infty} e^{-y^2/2}\,dy = 1 \qquad (5.3\text{-}7)$$

as was shown in Example 3.9.

The distribution function for the unit normal distribution is
given by

$$\Phi_X(x) = \frac{1}{\sqrt{2\pi}} \int_{-\infty}^{x} e^{-y^2/2}\,dy \qquad (5.3\text{-}8)$$

Since the integrand is even, it follows that

$$\Phi_X(-x) = 1 - \Phi_X(x) \qquad (5.3\text{-}9)$$

For the case $x = -x = 0$, this last expression becomes

$$\Phi_X(0) = 1/2 \qquad (5.3\text{-}10)$$

The integral of Eq. (5.3-8) cannot be evaluated analytically; therefore tables of $\Phi_X(x)$ (such as Table I) are obtained by some numerical approximation technique. Sometimes, instead of tables of the distribution function, one encounters tables of the *error integral* or *error function* $\Theta(y)$ where

$$\Theta(t) = erf \ t = \frac{2}{\sqrt{\pi}} \int_0^t e^{-y^2} \, dy \qquad (5.3\text{-}11)$$

With the substitution

$$y = \frac{x}{\sqrt{2}} \qquad (5.3\text{-}12)$$

Eq. (5.3-11) becomes

$$\Theta(t) = \frac{2}{\sqrt{2\pi}} \int_0^{\sqrt{2}\,t} e^{-x^2/2} \, dx \qquad (5.3\text{-}13)$$

It is clear from Eqs. (5.3-8) and (5.3-9) that the error function $\Theta(t)$ is related to the distribution function $\Phi_X(x)$ for the unit normal distribution by

$$\Theta(t) = 2\Phi_X(\sqrt{2}\,t) - 1 \qquad (5.3\text{-}14)$$

The unit normal distribution with density function $\phi_X(x)$ given by Eq. (5.3-4) has already been shown to have zero mean and unit variance. The normal density of Eq. (5.3-3) has mean m since

$$E(X) = \frac{1}{\sqrt{2\pi}\,\sigma} \int_{-\infty}^{\infty} x \ e^{-(x-m)^2/2\sigma^2} \, dx =$$

$$\frac{1}{\sqrt{2\pi}} \int_{-\infty}^{\infty} (\sigma y + m) e^{-y^2/2} \, dy \qquad (5.3\text{-}15)$$

or

$$E(X) = 0 + \frac{m}{\sqrt{2\pi}} \int_{-\infty}^{\infty} e^{-y^2/2} \, dy = m \qquad (5.3\text{-}16)$$

The variance is σ^2 and is found in the same way, that is,

$$E[(X-m)^2] = \frac{1}{\sqrt{2\pi}\,\sigma} \int_{-\infty}^{\infty} (x-m)^2 e^{-(x-m)^2/2\sigma^2} \, dx$$

$$= \frac{\sigma^2}{\sqrt{2\pi}} \int_{-\infty}^{\infty} y^2 \ e^{-y^2/2} \, dy = \sigma^2 \qquad (5.3\text{-}17)$$

The last integral was evaluated in Example 4.6.

The characteristic function $M_Y(iu)$ for a random variable Y with the unit normal distribution was found in Example 4.11 to be

$$M_Y(iu) = E(e^{iuY}) = e^{-u^2/2} \qquad (5.3\text{-}18)$$

The characteristic function $M_X(iu)$ for a random variable X with the normal distribution $N(m,\sigma^2)$ is

$$M_X(iu) = E(e^{iuX}) = \frac{1}{\sqrt{2\pi}} \int_{-\infty}^{\infty} e^{iux}\, e^{-(x-m)^2/2\sigma^2}\, dx \qquad (5.3\text{-}19)$$

This integral is typical of many integrals encountered when dealing with the normal distribution. The general procedure to follow is to complete the square in the variable of integration and then to make a linear change of variable. We have

$$M_X(iu) = e^{ium-u^2\sigma^2/2} \frac{1}{\sqrt{2\pi}\,\sigma} \int_{-\infty}^{\infty} e^{-\frac{|x-(m+iu\,\sigma^2)|^2}{2\sigma^2}} dx \qquad (5.3\text{-}20)$$

After the change of variable $y = [x-(m+iu\,\sigma^2)]/\sigma$, we obtain

$$M_X(iu) = e^{ium-u^2\sigma^2/2} \qquad (5.3\text{-}21)$$

for the characteristic function for the normal distribution. In the same way, the moment generating function $M_X(t)$ is found to be

$$M_X(t) = E(e^{tX}) = e^{tm+t^2\sigma^2/2} \qquad (5.3\text{-}22)$$

Another way to consider the problem is to recognize that a random variable X with the normal distribution $N(m,\sigma^2)$ is a linear transformation

$$X = \sigma Y + m \qquad (5.3\text{-}23)$$

of the random variable Y with the unit normal distribution $N(0,1)$. It follows from Eq. (4.6-4) that the characteristic function of $X = \sigma Y + m$ is

$$M_X(iu) = e^{ium}\, M_Y(\sigma iu) = e^{ium-u^2\sigma^2/2} \qquad (5.3\text{-}24)$$

in agreement with Eq. (5.3-21).

Usually, tables of only the unit normal distribution are available. For the normal distribution $N(m,\sigma^2)$, these tables may be used with the change of variable of Eq. (5.3-23).

Example 5.4

Find the distribution function $F_X(1.2)$ for the normal distribution $N(0.6, 9.0)$. We wish to find

$$F_X(1.2) = \frac{1}{\sqrt{2\pi}\,3} \int\limits_{-\infty}^{1.2} e^{-(x-0.6)^2/18}\, dx$$

We let

$$y = \frac{x-0.6}{3} \quad , \quad dx = 3\,dy$$

and obtain

$$F_X(1.2) = \frac{1}{\sqrt{2\pi}} \int\limits_{-\infty}^{0.2} e^{-y^2/2}\, dy = \Phi_Y(0.2)$$

where $\Phi_Y(y)$ is the distribution function of the unit normal distribution. It follows from Table I that

$$F_X(1.2) = \Phi_Y(0.2) = 0.5793$$

Consider the moment generating function $M_Y(t)$ for the unit normal distribution. It may be expanded as

$$M_Y(t) = e^{t^2/2} = 1 + \frac{t^2}{2} + \frac{1}{2!}\left(\frac{t^2}{2}\right)^2 +$$

$$+ \cdots + \frac{1}{n!}\left(\frac{t^2}{2}\right)^n + \cdots \qquad (5.3\text{-}25)$$

The n-th moment about the origin m_n is the coefficient of $\dfrac{t^n}{n!}$ in this power series expansion of $M_Y(t)$. There are no odd powers of t in Eq. (5.3-25) and, for n even, the $n/2$-th term is

$$\frac{1}{(n/2)!}\left(\frac{t^2}{2}\right)^{n/2} = \frac{n(n-1)...(\frac{n}{2}+1)}{2^{n/2}}\,\frac{t^n}{n!} \qquad (5.3\text{-}26)$$

Consequently, the moments about the origin for the unit normal distribution are

for n odd, $m_n = 0$

$$(5.3\text{-}27)$$

for n even, $m_n = \dfrac{n(n-1)...(\frac{n}{2}+1)}{2^{n/2}}$

Of course, since the mean is zero, m_n is also equal to u_n, the n-th moment about the mean.

5.4 *The Bivariate Normal Distribution* - Let us consider random variables X and Y with density functions $f_X(x)$ and $f_Y(y)$, respectively. As pointed out previously, a knowledge of these two densities is not sufficient in general to construct the joint density $f_{X,Y}(x,y)$. On the other hand, if X and Y are independent, we have immediately from Section 3.11

$$f_{X,Y}(x,y) = f_X(x)f_Y(y) \tag{5.4-1}$$

If X and Y are normal random variables with means m_X and m_Y and variances σ_X^2 and σ_Y^2, then this last expression becomes

$$f_{X,Y}(x,y) =$$

$$\tag{5.4-2}$$

$$\frac{1}{2\pi\sigma_X\sigma_Y} e^{-\frac{1}{2}[(\frac{x-m_X}{\sigma_X})^2 + (\frac{y-m_Y}{\sigma_Y})^2]} \quad , \quad -\infty < x,y < \infty$$

More generally, when X and Y are not independent, they are said to be *jointly normal* or *jointly Gaussian* when their joint density function $f_{X,Y}(x,y)$ is given by

$$f_{X,Y}(x,y) = C\, e^{-Q(x,y)} \quad , \quad -\infty < x,y < \infty \tag{5.4-3}$$

where the constant C is

$$C = \frac{1}{2\pi\sigma_X\sigma_Y\sqrt{1-\rho^2}} \tag{5.4-4}$$

and $Q(x,y)$ is the quadratic form

$$Q(x,y) =$$

$$\tag{5.4-5}$$

$$\frac{1}{2(1-\rho^2)}\left[\left(\frac{x-m_X}{\sigma_X}\right)^2 - 2\rho\left(\frac{x-m_X}{\sigma_X}\right)\left(\frac{y-m_Y}{\sigma_Y}\right) + \left(\frac{y-m_Y}{\sigma_Y}\right)^2\right]$$

It will turn out that m_X and m_Y are the means and σ_X^2 and σ_Y^2 are the variances of X and Y, respectively. The quantity ρ will be shown to be the correlation coefficient defined by Eq. (4.7-9); that is,

$$\rho = \rho_{X,Y} = \frac{\sigma_{X,Y}}{\sigma_X\sigma_Y} = \frac{E[(X-m_X)(Y-m_Y)]}{\sigma_X\sigma_Y} \tag{5.4-6}$$

Notice that, when $\rho=0$, Eq.(5.4-3) reduces to Eq. (5.4-2). Thus, normally distributed random variables which are *uncorrelated* are also *independent*. It was pointed out previously that this statement is not true for arbitrary random variables. For the case where $\rho=\pm1$, the expression given by Eq. (5.4-3) is not meaningful. As discussed in Section 4.7, in this case X and Y are linearly related and are said to have a *singular* normal distribution. The function $f_{X,Y}$ has non-zero values only on the line

$$\frac{x-m_X}{\sigma_X} = \pm \frac{y-m_Y}{\sigma_Y}$$

and the linear relationship between X and Y is

$$Y = \frac{\sigma_Y}{\sigma_X}(X-m_X) + m_Y \tag{5.4-7}$$

Let us now return to Eq. (5.4-3) and show that $f_{X,Y}(x,y)$ is in reality a density function. It is apparent from inspection that

$$f_{X,Y}(x,y) \geq 0 \ , \quad -\infty < x,y < \infty \tag{5.4-8}$$

It remains to be shown that $\int\limits_{-\infty}^{\infty} \int\limits_{-\infty}^{\infty} f_{X,Y}(x,y)dx \ dy = 1$. In the process of doing this, we shall also show that X and Y are individually normal random variables, that is, subject to univariate normal laws. The distribution function $F_X(x)$ of the random variable X is given by

$$F_X(x) = \int\limits_{-\infty}^{x} \int\limits_{-\infty}^{\infty} f_{X,Y}(x,y) \ dydx$$

$$= C \int\limits_{-\infty}^{x} \int\limits_{-\infty}^{\infty} e^{-Q(x,y)}dy \ dx \tag{5.4-9}$$

This expression may be rearranged to yield

$$F_X(x) =$$

$$C \int\limits_{-\infty}^{x} e^{-\frac{1}{2}\left[\frac{x-m_X}{\sigma_X}\right]^2} \int\limits_{-\infty}^{\infty} e^{-\frac{1}{2}\left[\frac{y-m_Y}{\sigma_Y\sqrt{1-\rho^2}} - \rho\frac{x-m_X}{\sigma_X\sqrt{1-\rho^2}}\right]^2} dy \ dx \tag{5.4-10}$$

We now consider the inner integral only, make the linear change of variable

$$z = \frac{y-m_Y}{\sigma_Y\sqrt{1-\rho^2}} - \rho \frac{x-m_X}{\sigma_X\sqrt{1-\rho^2}} \tag{5.4-11}$$

and obtain for this integral

$$\sigma_Y\sqrt{1-\rho^2} \int\limits_{-\infty}^{\infty} e^{-z^2/2} \ dz = \sqrt{2\pi}\,\sigma_Y\sqrt{1-\rho^2} \tag{5.4-12}$$

Now Eq. (5.4-10) becomes

$$F_X(x) = \frac{1}{\sqrt{2\pi}\sigma_X} \int\limits_{-\infty}^{\infty} e^{-\frac{1}{2}\left[\frac{x-m_X}{\sigma_X}\right]^2} dx \tag{5.4-13}$$

This is the distribution function for a random variable X which has the normal distribution $N(m_X, \sigma_X^2)$ discussed in Section 5.3. It has already been shown that $F_X(\infty) = 1$; hence it has been proved that

$$\int\limits_{-\infty}^{\infty} \int\limits_{-\infty}^{\infty} f_{X,Y}(x,y) \; dx \; dy \; = 1 \qquad (5.4\text{-}14)$$

It is clear from symmetry that a similar result would be obtained for $F_Y(y)$; that is; Y is normally distributed $N(m_Y, \sigma_Y^2)$.

We must now show that the constant ρ in Eq. (5.4-3) is the correlation coefficient given by Eq. (5.4-6). By direct substitution in Eq. (5.4-6), we obtain

$$\rho_{X,Y} = \frac{\sigma_{X,Y}}{\sigma_X \sigma_Y} \qquad (5.4\text{-}15)$$

$$= C \int\limits_{-\infty}^{\infty} \int\limits_{-\infty}^{\infty} \frac{x-m_X}{\sigma_X} \; \frac{y-m_Y}{\sigma_Y} \; e^{-Q(x,y)} \; dy \; dx$$

Rearrange this expression as in Eq. (5.4-10) and make the change of variable given by Eq. (5.4-11). The result is

$$\frac{\sigma_{X,Y}}{\sigma_X \sigma_Y} = C \int\limits_{-\infty}^{\infty} \frac{x-m_X}{\sigma_X} \; e^{-\frac{1}{2}\left[\frac{x-m_X}{\sigma_X}\right]^2} \times$$

$$(5.4\text{-}16)$$

$$\times \int\limits_{-\infty}^{\infty} \left[\sigma_Y(1-\rho^2)z + \sigma_Y\sqrt{1-\rho^2} \; \rho\left(\frac{x-m_X}{\sigma_X}\right)\right] e^{-z^2/2} \; dz dx$$

The first of the inner integrals is zero since the integrand is odd. The second inner integral becomes

$$\sigma_Y \sqrt{1-\rho^2} \; \rho \left(\frac{x-m_X}{\sigma_X}\right) \; \sqrt{2\pi}$$

since $\int\limits_{-\infty}^{\infty} e^{-z^2/2} \; dz = \sqrt{2\pi}$. Now Eq. (5.4-16) becomes

$$\rho_{X,Y} = \rho \; \frac{1}{\sqrt{2\pi} \; \sigma_X} \int\limits_{-\infty}^{\infty} \left(\frac{x-m_X}{\sigma_X}\right)^2 e^{-\frac{1}{2}\left[\frac{x-m_X}{\sigma_X}\right]^2} \; dx \qquad (5.4\text{-}17)$$

Again a linear change of variable $w = (x-m_X)/\sigma_X$ yields

$$\rho_{X,Y} = \rho \; \frac{1}{\sqrt{2\pi}} \int\limits_{-\infty}^{\infty} w^2 \; e^{-w^2/2} \; dw = \rho \qquad (5.4\text{-}18)$$

Thus, as postulated, the constant ρ is the correlation coefficient between X and Y.

Both the joint characteristic function $M_{X,Y}(u,v)$ and the joint moment generating function $M_{X,Y}(t,s)$ are easily obtained for the random variables X and Y with the bivariate normal distribution. In accordance with Eq. (4.6-7), the joint characteristic function is

$$M_{X,Y}(iu,iv) = E(e^{i(uX+vY)}) \tag{5.4-19}$$

or, from Eq. (5.4-3),

$$M_{X,Y}(iu,iv) = C \int_{-\infty}^{\infty} \int_{-\infty}^{\infty} e^{i(ux+vy)} e^{-Q(x,y)} \, dy \, dx \tag{5.4-20}$$

The computation is lengthy but straightforward. Consider the exponent only of the integrand. This exponent will be called N and may be written as

$$N = iux + ivy$$

$$\tag{5.4-21}$$

$$- \frac{1}{2(1-\rho^2)} \left[\left(\frac{x-m_X}{\sigma_X} \right)^2 - 2\rho \left(\frac{x-m_X}{\sigma_X} \right) \left(\frac{y-m_Y}{\sigma_Y} \right) + \left(\frac{y-m_Y}{\sigma_Y} \right)^2 \right]$$

Let us begin with the change in variables

$$t = \frac{x-m_X}{\sigma_X} \quad, \quad dx = \sigma_X \, dt \tag{5.4-22}$$

$$s = \frac{y-m_Y}{\sigma_Y} \quad, \quad dy = \sigma_Y \, ds \tag{5.4-23}$$

Now the exponent becomes

$$N = N(t,s) =$$

$$iu(\sigma_X t + m_X) + iv(\sigma_Y s + m_Y) - \frac{1}{2(1-\rho^2)} [t^2 - 2\rho ts + s^2] \tag{5.4-24}$$

and Eq. (5.4-20) becomes

$$M_{X,Y}(iu,iv) = \frac{1}{2\pi\sqrt{1-\rho^2}} \int_{-\infty}^{\infty} \int_{-\infty}^{\infty} e^{N(t,s)} dt ds \tag{5.4-25}$$

As before, we complete the square in the variable s so that the exponent $N(t,s)$ is rearranged as follows:

$$N(t,s) = i(um_X + vm_Y) + iu\,\sigma_X t - \frac{t^2}{2} + iv\,\sigma_Y s - \frac{(s-\rho t)^2}{2(1-\rho^2)} \tag{5.4-26}$$

Considering first the integral with respect to s in Eq. (5.4-25) we have

$$\frac{1}{\sqrt{2\pi(1-\rho^2)}} \int_{-\infty}^{\infty} e^{iv\,\sigma_Y s} \, e^{-\frac{(s-\rho t)^2}{2(1-\rho^2)}} \, ds$$

A comparison of this integral with Eq. (5.3-19) shows that it is the characteristic function of a normally distributed random variable with mean ρt and variance $(1-\rho^2)$ and with the dummy variable u replaced by $v\,\sigma_Y$. Thus the value of this integral is

$$e^{iv\,\sigma_Y \rho t - v^2 \sigma_Y^2 (1-\rho^2)/2}$$

Now Eq. (3.4-25) becomes

$$M_{X,Y}(iu,iv) =$$

$$e^{i(um_X+vm_Y)-v^2\sigma_Y^2(1-\rho^2)/2}\, \frac{1}{\sqrt{2\pi}}\int_{-\infty}^{\infty} e^{H(t)}\, dt \tag{5.4-27}$$

where $H(t)$ may be written as

$$H(t) = i(u\,\sigma_X+v\,\sigma_Y\rho)t - t^2/2 \tag{5.4-28}$$

Again the integral $\dfrac{1}{\sqrt{2\pi}}\displaystyle\int_{-\infty}^{\infty} e^{H(t)}\, dt$ is the characteristic function of a normally distributed random variable with mean zero, variance of unity and with the dummy variable u replaced by $(u\,\sigma_X+v\,\sigma_Y\rho)$. It follows that

$$\frac{1}{\sqrt{2\pi}}\int_{-\infty}^{\infty} e^{H(t)}\, dt = e^{-(u\,\sigma_X+v\,\sigma_Y\rho)^2/2} \tag{5.4-29}$$

Finally, on putting all of this together, we find from Eq. (5.4-27) that the joint characteristic function for the bivariate normal distribution is

$$M_{X,Y}(iu,iv) =$$

$$e^{i(um_X+vm_Y)-\frac{1}{2}[u^2\sigma_X^2+2uv\,\rho\sigma_X\sigma_Y+v^2\sigma_Y^2]} \tag{5.4-30}$$

As a matter of interest, note that the marginal distributions can be found easily from this last expression. For example, on setting $v=0$, we have the characteristic function for X; that is,

$$M_{X,Y}(iu,0) = e^{ium_X-u^2\sigma_X^2/2} \tag{5.4-31}$$

This is the characteristic function of a normally distributed random variable with mean m_X and variance σ_X^2. Notice also that when $\rho=0$, then X and Y are independent random variables and, as expected,

$$M_{X,Y}(iu,iv) = M_X(iu)M_Y(iv) \;,\; \rho=0 \tag{5.4-32}$$

Let X and Y be independent random variables which are normally distributed $N(m_X,\sigma_X^2)$ and $N(m_Y,\sigma_Y^2)$, respectively. Consider the random variable Z defined by

$$Z = X + Y \tag{5.4-33}$$

with characteristic function

$$M_Z(iu) = E(e^{iuZ}) = E(e^{iuX})E(e^{iuY}) \tag{5.4-34}$$

It follows from Eq. (5.3-21) that M_Z is

$$M_Z(iu) = e^{iu(m_X+m_Y)-u^2(\sigma_X^2+\sigma_Y^2)/2} \tag{5.4-35}$$

Thus the sum of two independent normal random variables is itself a normal random variable distributed $N(m_X+m_Y,\sigma_X^2+\sigma_Y^2)$. More generally, let X and Y be jointly normal with joint density function $f_{X,Y}$ given by Eq. (5.4-3) and joint characteristic function $M_{X,Y}$ given by Eq. (5.4-30). It is clear that the characteristic function of $Z = X + Y$ is just $M_{X,Y}$ with $u=v$, that is

$$M_Z(iu) = e^{iu(m_X+m_Y)-\frac{u^2}{2}[\sigma_X^2+2\rho\sigma_X\sigma_Y+\sigma_Y^2]} \qquad (5.4\text{-}36)$$

In other words, the sum of any two jointly normal random variables X and Y is normal with mean m_X+m_Y and variance $\sigma_X^2+2\rho\sigma_X\sigma_Y+\sigma_Y^2$.

When X and Y are independent random variables which are normally distributed and which have equal variances σ^2, they are said to be *circularly normal*. In this case, their joint density function is simply given by

$$f_{X,Y}(x,y) = \frac{1}{2\pi\sigma^2}\, e^{-\frac{(x-m_X)^2+(y-m_Y)^2}{2\sigma^2}} \qquad , \quad -\infty<x,y<\infty \quad (5.4\text{-}37)$$

On account of the symmetry involved, this density function may be somewhat easier to manipulate.

5.5 *Rotations for Independence* - The bivariate density $f_{X,Y}$ for two jointly normal random variables X and Y is given by Eq. (5.4-3). For jointly normal variables it is always possible to find a linear transformation on these random variables that will yield two *independent* normal random variables.

It is convenient to begin by centering X and Y through the transformations

$$X_1 = X - m_X \qquad (5.5\text{-}1)$$

$$X_2 = Y - m_Y \qquad (5.5\text{-}2)$$

The quadratic form $Q(x,y)$ of Eq. (3.5-5) now becomes

$$Q(x_1,x_2) = \qquad (5.5\text{-}3)$$

$$\frac{1}{2(1-\rho^2)}\left[\left(\frac{x_1}{\sigma_1}\right)^2 - 2\rho\left(\frac{x_1}{\sigma_1}\right)\left(\frac{x_2}{\sigma_2}\right) + \left(\frac{x_2}{\sigma_2}\right)^2\right]$$

since $\quad Var(X_1) = \sigma_1^2 = Var(X) = \sigma_X^2 \quad$ and $Var(X_2) = \sigma_2^2 = Var(Y) = \sigma_Y^2$. The density function $f_{X_1,X_2}(x_1,x_2)$ can be written as

$$f_{X_1,X_2}(x_1,x_2) = C\, e^{-Q(x_1,x_2)} \qquad (5.5\text{-}4)$$

where C has been defined previously by Eq. (3.4-4). The correlation coefficient ρ is

$$\rho = \frac{Cov(X,Y)}{\sigma_X \sigma_Y} = \frac{Cov(X_1,X_2)}{\sigma_1 \sigma_2} = \frac{E(X_1 X_2)}{\sigma_1 \sigma_2} \qquad (5.5\text{-}5)$$

The translations of Eqs. (5.5-1) and (5.5-2) have simply shifted the means to the origin without affecting the central moments.

Let us now make the *linear* change of random variables expressed by the equations

$$Z_1 = X_1 - \rho \frac{\sigma_1}{\sigma_2} Z_2 \qquad (5.5\text{-}6)$$

where

$$Z_2 = X_2 \qquad (5.5\text{-}7)$$

As pointed out previously, the random variable Z_1 is normal since it is a linear combination of two jointly normal random variables X_1 and X_2. The transformation to Z_2 is simply a formal transformation for notational convenience, that is, Z_2 is X_2, which is a normal random variable. It has not been shown that Z_1 and Z_2 are jointly normal, but this will be done shortly; let us accept its validity for the moment. Now if Z_1 and Z_2 are uncorrelated, they are independent, since they are jointly normal. Note that $E(Z_1) = E(Z_2) = 0$; hence the covariance of Z_1 and Z_2 is given by

$$Cov(Z_1 Z_2) = E(Z_1 Z_2) = E(X_1 X_2) + \rho \frac{\sigma_1}{\sigma_2} E(X_2^2) \qquad (5.5\text{-}8)$$

It follows from Eq. (5.5-5) that

$$Cov(Z_1 Z_2) = \rho \sigma_1 \sigma_2 - \rho \sigma_1 \sigma_2 = 0 \qquad (5.5\text{-}9)$$

In other words, on the assumption that Z_1 and Z_2 are jointly normal, they are independent since uncorrelated normal random variables are independent. The linear transformation (rotation) given by Eqs. (5.5-6) and (5.5-7) converts the jointly normal random variables X_1 and X_2 with correlation coefficient ρ into independent normal random variables Z_1 and Z_2. Furthermore the variances of Z_1 and Z_2 follow immediately from Eqs. (5.5-6) and (5.5-7)

$$Var(Z_1) = E(Z_1^2) =$$

$$E(X_1^2) - 2\rho \frac{\sigma_1}{\sigma_2} E(X_1 X_2) + \rho^2 \frac{\sigma_1^2}{\sigma_2^2} E(X_2^2) = \sigma_1^2(1-\rho^2) \quad (5.5\text{-}10)$$

and

$$Var(Z_2) = E(X_2^2) = \sigma_2^2 \qquad (5.5\text{-}11)$$

Let x_1 and x_2 be replaced by z_1 and z_2 in Eq. (5.5-3). The result is that $Q(x_1,x_2)$ becomes $Q(z_1,z_2)$ where

$$Q(z_1, z_2) = \frac{1}{2} \left[\frac{z_1^2}{\sigma_1^2(1-\rho^2)} + \frac{z_2^2}{\sigma_2^2} \right] \tag{5.5-12}$$

Consider now what happens to the joint distribution function $F_{X_1, X_2}(a, b)$ where

$$F_{X_1, X_2}(a, b) = C \int_{-\infty}^{b} \int_{-\infty}^{a} e^{-Q(x_1, x_2)} \, dx_1 \, dx_2 \tag{5.5-13}$$

We now make the change of variables in this definite integral as given by Eqs. (5.5-6) and (5.5-7). The differential volume $dx_1 dx_2$ becomes

$$dx_1 dx_2 = J\left(\frac{x_1, x_2}{z_1, z_2} \right) dz_1 dz_2 \tag{5.5-14}$$

where J is the Jacobian of the transformation and is given by

$$J\left(\frac{x_1, x_2}{z_1, z_2} \right) = \begin{vmatrix} \dfrac{\partial x_1}{\partial z_1} & \dfrac{\partial x_1}{\partial z_2} \\ \dfrac{\partial x_2}{\partial z_1} & \dfrac{\partial x_2}{\partial z_2} \end{vmatrix} = \begin{vmatrix} 1 & \rho\dfrac{\sigma_1}{\sigma_2} \\ 0 & 1 \end{vmatrix} = 1 \tag{5.5-15}$$

Thus Eq. (5.5-14) becomes

$$F_{X_1, X_2}(a, b) = C \int_{-\infty}^{b} \int_{-\infty}^{a - \rho\frac{\sigma_1}{\sigma_2} b} e^{-Q(z_1, z_2)} \, dz_1 \, dz_2 \tag{5.5-16}$$

On using Eq. (5.4-4) and (5.5-12) we may rewrite this last equation as

$$F_{X_1, X_2}(a, b) = \left[\frac{1}{\sqrt{2\pi}\sigma_1 \sqrt{1-\rho^2}} \int_{-\infty}^{a - \rho\frac{\sigma_1}{\sigma_2} b} e^{-z_1^2/2\sigma_1^2(1-\rho^2)} \, dz_1 \right] \times$$

$$\left[\frac{1}{\sqrt{2\pi}\sigma_2} \int_{-\infty}^{b} e^{-z_2^2/2\sigma_2^2} \, dz_2 \right] \tag{5.5-17}$$

or

$$F_{X_1, X_2}(a, b) = F_{Z_1}\left(a - \rho\frac{\sigma_1}{\sigma_2} b\right) F_{Z_2}(b) \tag{5.5-18}$$

It follows from the definition of distribution function that

$$F_{X_1,X_2}(a,b) = P\{X_1 \le a \text{ and } X_2 \le b\}$$

$$= P\{Z_1 \le a - \rho \frac{\sigma_1}{\sigma_2} b \text{ and } Z_2 \le b\} \quad (5.5\text{-}19)$$

$$= F_{Z_1,Z_2}\left(a - \rho \frac{\sigma_1}{\sigma_2} b, b\right)$$

The joint distribution function F_{Z_1,Z_2} factors into the product of the two univariate distribution functions F_{z_1} and F_{z_2}. It is clear that Z_1 and Z_2 are independent normal random variables with zero means and variances $\sigma_1^2(1-\rho^2)$ and σ_2^2, respectively. This result could be derived also by finding the joint characteristic function of Z_1 and Z_2.

PROBLEMS

1. When X is binomially distributed, show that

$$P\{X = even\} = \frac{1}{2}[1 + (q-p)^n] \ , \quad q = 1-p$$

2. When X is Poisson distributed, show that

$$P\{X = even\} = \frac{1}{2}[1 + e^{-2\lambda}]$$

3. Show that a recursion formula for the binomial distribution $b(k;n,p)$ is

$$P\{X = k+1\} = \frac{p(n-k)}{q(k+1)} P\{X = k\} \ , \quad k = 0,1,...,n-1$$

   where $q = 1-p$.

4. Show that a recursion formula for the Poisson distribution $p(k,\lambda)$ is

$$P\{X = k+1] = \frac{\lambda}{k+1} P\{X = k\} \ , \quad k = 0,1,2,...$$

5. In Example 5.3, use Stirling's formula to evaluate the probability $b(20;40,1/2)$.

6. For $\lambda = 20$, plot approximately the Poisson density function

$$p(k,\lambda) = e^{-\lambda} \frac{\lambda^k}{k!}$$

7. For the *circularly normal* density function of Eq. (5.4-37), find the probability that the point (X,Y) will lie within a circle of radius r_1 centered on the point (m_X, m_Y).

8. Consider the integral

$$A(x) = \frac{1}{\sqrt{2\pi}} \int_0^x e^{-y^2/2} \, dy$$

By considering A^2 as an integral in the plane, prove that

$$\frac{1}{4} \left(1 - e^{-x^2/2}\right) < A^2(x) < \frac{1}{4} \left(1 - e^{x^2}\right)$$

9. A fair coin is tossed ten times. Find the probability that the number of heads is 4, 5, or 6 and also find the combined probability of these three events, using (a) the binomial distribution and (b) the normal approximation.

10. A fair coin is tossed 200 times. Find the combined probability that a head will appear 99, 100, or 101 times.

11. In a book 200 pages long, it is not unreasonable to expect 100 misprints (fortunately, most of them are obvious). Find the probability that a given page will contain (a) two misprints, (b) two or less misprints, and (c) two or more misprints.

Chapter 6

THE MULTIVARIATE NORMAL DISTRIBUTION

In this chapter we generalize the normal distribution to the multivariate case. The multivariate normal or Gaussian distribution is one of the most important multivariate distributions encountered in applied probability. One of the reasons is that this distribution is fundamental in the definition of a normal or Gaussian random process. A random process is a time- varying random variable; i.e., a random waveform. The Gaussian random process arises in a multitude of applications, both because it is a good model for many kinds of physical noises and random disturbances and because it is one of the most tractable of the non-trivial random processes.

It will be convenient to introduce some elementary matrix concepts in order to simplify the notation. For the reader who is unfamiliar with these, the barest outlines are given in Appendix D. As we proceed we will often use the bivariate normal distribution of Section 5.4 as an illustration.

6.1 *The Covariance Matrix* - In previous developments we have used the notation $\mathbf{X}_n = (X_1, X_2, ..., X_n)$ to denote a set of n random variables. Henceforth the symbol $\mathbf{X}$ will be used to denote the column vector (or n × one matrix) of n random variables; in other words

$$\mathbf{X} = \begin{bmatrix} X_1 \\ X_2 \\ \cdot \\ \cdot \\ \cdot \\ X_n \end{bmatrix} \tag{6.1-1}$$

The transpose of $\mathbf{X}$ will be denoted by $\mathbf{X}^T$ and is the row vector (or one × n matrix)

$$\mathbf{X}^T = [X_1, X_2, ..., X_n] \tag{6.1-2}$$

The *covariance matrix* Λ of $\mathbf{X}$ is the $n \times n$ matrix

$$\Lambda = [\sigma_{ij}] = \begin{bmatrix} \sigma_{11} & \sigma_{21} & \cdots & \sigma_{1n} \\ \sigma_{21} & \sigma_{22} & \cdots & \sigma_{2n} \\ \cdot & \cdot & & \cdot \\ \cdot & \cdot & & \cdot \\ \cdot & \cdot & & \cdot \\ \sigma_{n1} & \sigma_{n2} & \cdots & \sigma_{nn} \end{bmatrix} \qquad (6.1\text{-}3)$$

where the element σ_{jk} is the covariance of the random variables X_j and X_k; that is

$$\sigma_{jk} = Cov(X_j, X_k) = E[(X_j - m_j)(X_k - m_k)] \qquad (6.1\text{-}4)$$

and $m_j = E(X_j)$ and $m_k = E(X_k)$ are the means of X_j and X_k, respectively. Note that

$$\sigma_{jk} = \sigma_{kj} \qquad (6.1\text{-}5)$$

so that the covariance matrix Λ is *symmetrical*. The determinant of Λ will be denoted by $|\Lambda|$ or $|\sigma_{ij}|$.

A column vector **m** of the means of the X_i will be given by

$$\mathbf{m} = \begin{bmatrix} m_1 \\ m_2 \\ \cdot \\ \cdot \\ \cdot \\ m_n \end{bmatrix} \qquad (6.1\text{-}6)$$

where $m_i = E(X_i)$. Now the covariance matrix Λ can be written as*

$$\Lambda = E[(\mathbf{X} - \mathbf{m})(\mathbf{X} - \mathbf{m})^T] \qquad (6.1\text{-}7)$$

Let us now define an arbitrary but non-zero column vector **u** of real variables u_i by

$$\mathbf{u} = \begin{bmatrix} u_1 \\ u_2 \\ \cdot \\ \cdot \\ \cdot \\ u_n \end{bmatrix} \qquad (6.1\text{-}8)$$

---

*By the expectation of a matrix of random variables, we mean the matrix whose jk-th element is the expectation of the jk-th element of the random matrix.

The *quadratic form* associated with the covariance matrix Λ can be written as

$$\mathbf{u}^T \Lambda \mathbf{u} = \sum_{j=1}^{n} \sum_{k=1}^{n} u_j \, \sigma_{jk} \, u_k \tag{6.1-9}$$

where σ_{jk} is the jk-th element of Λ. Let the random variable Y be defined by

$$Y = \sum_{j=1}^{n} u_j (X_j - m_j) \tag{6.1-10}$$

It is clear that the quantity $E(Y^2)$ is non-negative and can be written as

$$E(Y^2) = E\left[\sum_{j=1}^{n} u_j (X_j - m_j) \sum_{k=1}^{n} u_k (X_k - m_k) \right] \geq 0 \tag{6.1-11}$$

This expression may be rearranged and Eqs. (6.1-4) and (6.1-9) may be used to yield

$$\sum_{j=1}^{n} \sum_{k=1}^{n} u_j \, E\left[(X_j - m_j)(X_k - m_k) \right] u_k = \sum_{j=1}^{n} \sum_{k=1}^{n} u_j \, \sigma_{jk} \, u_k$$

$$= \mathbf{u}^T \Lambda \mathbf{u} \geq 0 \tag{6.1-12}$$

Thus the covariance matrix of any random vector is *non-negative definite*. Actually, we shall avoid the degenerate case where the equality holds; that is, where $E(Y^2) = 0$, and consider only column vectors $\mathbf{X}$ such that their covariance matrices are *positive definite;* that is,

$$\mathbf{u}^T \Lambda \mathbf{u} > 0 \tag{6.1-13}$$

for arbitrary $\mathbf{u}$.

The inverse of the covariance matrix Λ will be denoted by

$$\Lambda^{-1} = [\sigma^{ij}] = \begin{bmatrix} \sigma^{11} & \sigma^{12} & \cdots & \sigma^{1n} \\ \sigma^{21} & \sigma^{22} & \cdots & \sigma^{2n} \\ \cdot & \cdot & & \cdot \\ \cdot & \cdot & & \cdot \\ \cdot & \cdot & & \cdot \\ \sigma^{n1} & \sigma^{n2} & \cdots & \sigma^{nn} \end{bmatrix} \tag{6.1-14}$$

with determinant*

$$|\Lambda^{-1}| = |\sigma^{ij}| = \frac{1}{|\Lambda|} = \frac{1}{|\sigma_{ij}|} \tag{6.1-15}$$

The characteristic function of $\mathbf{X}$ will be written as

---

*Note that all positive definite matrices are invertible.

$$M_X(i\ \mathbf{u}) = E(e^{i\ \mathbf{u}^T\ \mathbf{X}}) = E[exp\,(i\ \sum_{j=1}^{n} u_j\ X_j)] \qquad (6.1\text{-}16)$$

where the column vector $\mathbf{u}$ has already been defined. Obviously, the moment generating function of $\mathbf{X}$ is

$$M_X(t) = E(e^{t^T \mathbf{X}}) \qquad (6.1\text{-}17)$$

where $\mathbf{t}$ is an n-element column vector with real elements t_i,

$$\mathbf{t} = \begin{bmatrix} t_1 \\ t_2 \\ . \\ . \\ . \\ t_n \end{bmatrix} \qquad (6.1\text{-}18)$$

6.2 *The Bivariate Normal Distribution in Matrix Form* - It will be instructive to use the matrix notation just developed to represent the bivariate normal density function given by Eq. (5.4-3) and associated Eqs. (5.4-4) and (5.4-5). The random variable X will be replaced by X_1 and Y by X_2.

Note that the covariance matrix Λ can be written as

$$\Lambda = \begin{bmatrix} \sigma_{11} & \sigma_{12} \\ \sigma_{21} & \sigma_{22} \end{bmatrix} = \begin{bmatrix} \sigma_1^2 & \sigma_1\sigma_2\rho \\ \sigma_1\sigma_2\rho & \sigma_2^2 \end{bmatrix} \qquad (6.2\text{-}1)$$

Here ρ is the correlation coefficient of Eq. (5.4-6). An inconsistency in notation also arises since the variances of X_1 and X_2 have been written in two different ways, that is,

$$\sigma_{11} = \sigma_1^2 = E\,[(X_1\text{-}m_1)^2] \qquad (6.2\text{-}2)$$

and

$$\sigma_{22} = \sigma_2^2 = E\,[(X_2\text{-}m_2)^2] \qquad (6.2\text{-}3)$$

The determinant of Λ is

$$|\,\Lambda\,| = |\,\sigma_{ij}\,| = \sigma_1^2\sigma_2^2(1\text{-}\rho^2) \qquad (6.2\text{-}4)$$

The inverse of Λ is

$$\Lambda^{-1} = [\sigma^{ij}] = \begin{bmatrix} \dfrac{\sigma_2^2}{|\,\sigma_{ij}\,|} & -\dfrac{\sigma_1\sigma_2\rho}{|\,\sigma_{ij}\,|} \\ -\dfrac{\sigma_1\sigma_2\rho}{|\,\sigma_{ij}\,|} & \dfrac{\sigma_1^2}{|\,\sigma_{ij}\,|} \end{bmatrix} \qquad (6.2\text{-}5)$$

or

$$\mathbf{\Lambda}^{-1} = [\sigma^{ij}] = \begin{bmatrix} \dfrac{1}{\sigma_1^2(1-\rho^2)} & \dfrac{-\rho}{\sigma_1\sigma_2(1-\rho^2)} \\ \dfrac{-\rho}{\sigma_1\sigma_2(1-\rho^2)} & \dfrac{1}{\sigma_2^2(1-\rho^2)} \end{bmatrix} \tag{6.2-6}$$

with determinant

$$|\sigma^{ij}| = \frac{1}{|\sigma_{ij}|} = \frac{1}{\sigma_1^2\sigma_2^2(1-\rho^2)} \tag{6.2-7}$$

Consider now the quadratic form given by Eq. (5.4-5) which, in the present notation, can be written as

$$Q(x_1, x_2) = -\frac{1}{2}(\mathbf{x}-\mathbf{m})^T \mathbf{\Lambda}^{-1} (\mathbf{x}-\mathbf{m}) \tag{6.2-8}$$

where $\mathbf{m}$ is the column vector of means

$$\mathbf{m} = \begin{bmatrix} E(X_1) \\ E(X_2) \end{bmatrix} = \begin{bmatrix} m_1 \\ m_2 \end{bmatrix} \tag{6.2-9}$$

and $\mathbf{x}$ is another column vector given by

$$\mathbf{x} = \begin{bmatrix} x_1 \\ x_2 \end{bmatrix} \tag{6.2-10}$$

The constant C of Eq. (5.4-4) becomes

$$C = \frac{|\sigma^{ij}|^{1/2}}{2\pi} \tag{6.2-11}$$

Finally, the bivariate normal density function can be written as

$$f_{\mathbf{X}}(\mathbf{x}) = \frac{|\sigma^{ij}|^{1/2}}{2\pi} e^{-\frac{1}{2}(\mathbf{x}-\mathbf{m})^T \mathbf{\Lambda}^{-1} (\mathbf{x}-\mathbf{m})} \tag{6.2-12}$$

corresponding to Eq. (5.4-3).

In the same way, a comparison of Eqs. (5.4-30) and (6.1-16) shows that the characteristic function for the bivariate normal distribution can be written specifically as

$$M_{\mathbf{X}}(i\,\mathbf{u}) = e^{i\,\mathbf{m}^T\mathbf{u} - \mathbf{u}^T\mathbf{\Lambda}\mathbf{u}/2} \tag{6.2-13}$$

The moment generating function of Eq. (6.1-17) becomes

$$M_{\mathbf{X}}(\mathbf{t}) = e^{\mathbf{m}^T\mathbf{t} + \mathbf{t}^T\mathbf{\Lambda}\mathbf{t}/2} \tag{6.2-14}$$

6.3 *The Multivariate Normal Distribution* - The matrix notation of the previous section may be used to define the multivariate normal distribution and, in the process, to write the density function, the characteristic function, and the moment generating function for n jointly normal random variables. The form of the characteristic function and of the moment generating function is taken to be exactly the same as in the bivariate case. Thus the characteristic function is just Eq. (6.2-13)

$$M_X(i\ \mathbf{u}) = e^{\ i\ \mathbf{m}^T\ \mathbf{u}\ -\ \mathbf{u}^T\ \Lambda\ \mathbf{u}/2} \qquad (6.3\text{-}1)$$

where the row vectors $\mathbf{X}$, $\mathbf{m}$ and $\mathbf{u}$ have n components and the covariance matrix Λ is an $n \times n$ matrix. The exponents of Eq. (6.3-1) may be written in alternative form as

$$i\ \mathbf{m}^T\ \mathbf{u} = i\ \sum_{j=1}^{n} m_j\ u_j \qquad (6.3\text{-}2)$$

and

$$-\frac{1}{2}\ \mathbf{u}^T\ \Lambda\ \mathbf{u} = -\frac{1}{2}\ \sum_{j=1}^{n} \sum_{k=1}^{n} \sigma_{jk}\ u_j\ u_k \qquad (6.3\text{-}3)$$

In the same way, the characteristic function for the n-variate normal case is Eq. (6.2-14)

$$M_X(\mathbf{t}) = e^{\ \mathbf{m}^T \mathbf{t}\ +\ \mathbf{t}^T\ \Lambda\ \mathbf{t}/2} \qquad (6.3\text{-}4)$$

The n-variate normal density could be found by taking the multi-dimensional Fourier transform of the characteristic function using a generalization of Eq. (4.6-12). We could write

$$f_X(\mathbf{x}) = \frac{1}{(2\pi)^n} \int_{-\infty}^{\infty} \cdots \int_{-\infty}^{\infty} e^{-i\mathbf{u}^T \mathbf{x}} M_X(i\ \mathbf{u}) du_1 \ldots du_n \qquad (6.3\text{-}5)$$

If this complicated procedure is carried out, the result is that the n-variate density $f_X(\mathbf{x})$ differs slightly from the form given by Eq. (6.2-12) for the bivariate case and is

$$f_X(\mathbf{x}) = \frac{|\ \sigma^{ij}\ |^{1/2}}{(2\pi)^{n/2}}\ e^{-(\mathbf{x}-\mathbf{m})^T\ \Lambda^{-1}\ (\mathbf{x}-\mathbf{m})/2} \qquad (6.3\text{-}6)$$

Note that, in this expression, only the normalizing constant $(2\pi)^{-n/2}$ varies in *form* with the number of jointly normal random variables involved.

Most of the details have been omitted in arriving at the multivariate normal density $f_X(\mathbf{x})$ of Eq. (6.3-6). It has not been shown, for example that $f_X(\mathbf{x})$ is a legitimate density or, equivalently, that $M_X(i\ \mathbf{u})$ is a legitimate characteristic function. Some of these oversights will be corrected later. It is appropriate, however, to comment further at this time about the covariance matrix Λ and its inverse Λ^{-1} which determine the multivariate normal distribution. It has already been shown in Section 6.1 that the covariance matrix of any random vector $\mathbf{X}$ is non-negative definite and, except in certain degenerate cases, is actually positive definite as defined by Eq. (6.1-13). We shall say that Eq. (6.3-6) is a nondegenerate (or nonsingular) n-variate normal density function if and only if the quadratic form associated with the covariance matrix Λ is positive definite; that is, if Eq. (6.1-13) holds for arbitrary, but non-zero, $\mathbf{u}$. As pointed out in Appendix D, the inverse matrix Λ^{-1} then exists and the quadratic form associated with it is positive definite.

The degenerate or singular case where the quadratic form associated with Λ may be zero for some non-zero $\mathbf{u}$ has been mentioned before. It has been shown by Cramer (1946) that such a case corresponds to a situation where at least one of the n random variables X_i is a linear combination of others of the $n-1$ random variables.

An examination of Eq. (6.1-13) does not provide obvious clues as to how to construct a matrix Λ which is positive definite. In the bivariate case, the matrix is given by Eq. (6.2-1). In this simple situation the only requirements are that $\sigma_{11} > 0$, $\sigma_{22} > 0$, and $|\rho_{12}| < 1$. In general, however, it is not sufficient that the main diagonal terms (the variances) be positive and that the various correlation coefficients ρ_{ij} lie between -1 and $+1$.

It has been shown by Hohn (1973) that the following conditions on an $n \times n$ matrix Λ are necessary and sufficient for the quadratic form associated with it to be positive definite. The $k-th$ *leading principle minor* of the matrix Λ is the determinant D_k defined by

$$D_k = \begin{vmatrix} \sigma_{11} & \sigma_{12} & \cdots & \sigma_{1k} \\ \sigma_{21} & \sigma_{22} & \cdots & \sigma_{2k} \\ \cdot & & & \\ \cdot & & & \\ \cdot & & & \\ \sigma_{k1} & \sigma_{k2} & \cdots & \sigma_{kk} \end{vmatrix} \tag{6.3-7}$$

where the matrix Λ has elements σ_{ij} as defined by Eq. (6.1-13). The quadratic form associated with Λ is positive definite if and only if

$$D_k > 0 \quad , \quad k=1,2,...,n \tag{6.3-8}$$

It is quite clear that D_1 and D_2 are easily made to satisfy these requirements since

$$D_1 = \sigma_{11} = \sigma_1^2 > 0 \tag{6.3-9}$$

if the variance of X_1 is non-zero. Also

$$D_2 = \sigma_{11}\sigma_{22}(1-\rho_{12}^2) > 0 \tag{6.3-10}$$

if $\sigma_{11} > 0$, $\sigma_{22} > 0$ and $|\rho_{12}| < 1$. For values of n much greater than 2 the checking may be straightforward in principal but tedious in practice.

Let us now consider an alternative method of obtaining the multivariate normal density function of Eq. (6.3-6). Suppose that $Y_1, Y_2, ..., Y_n$ is a set of *independent* identically distributed *unit* normal random variables which will be represented by the column vector $\mathbf{Y}$ where

$$\mathbf{Y} = \begin{bmatrix} Y_1 \\ Y_2 \\ \cdot \\ \cdot \\ \cdot \\ Y_n \end{bmatrix} \qquad (6.3\text{-}11)$$

Thus the density function of each Y_i is given, for $-\infty < y_i < \infty$, by

$$f_{Y_i}(y_i) = \frac{1}{\sqrt{2\pi}} e^{-y_i^2/2} \quad , i = 1, 2, ..., n , \qquad (6.3\text{-}12)$$

As discussed in Section 3.11, the joint density $f_{\mathbf{Y}}(\mathbf{y})$ is just the product of the individual densities or

$$f_{\mathbf{Y}}(\mathbf{y}) = \prod_{i=1}^{n} f_{Y_i}(y_i) = \frac{1}{(2\pi)^{n/2}} e^{-\sum_{i=1}^{n} y_i^2/2}$$

$$= \frac{1}{(2\pi)^{n/2}} e^{-\mathbf{y}^T \mathbf{y}/2} \qquad (6.3\text{-}13)$$

where $\mathbf{y}$ is the column vector

$$\mathbf{y} = \begin{bmatrix} y_1 \\ y_2 \\ \cdot \\ \cdot \\ \cdot \\ y_n \end{bmatrix} \qquad (6.3\text{-}14)$$

Consider now the general linear transformation

$$\mathbf{X} = \mathbf{A}\,\mathbf{Y} + \mathbf{b} \qquad (6.3\text{-}15)$$

where $\mathbf{A}$ is an $n \times n$ matrix with constant elements a_{ij} so that

$$\mathbf{A} = \begin{bmatrix} a_{11} & a_{12} & \cdots & a_{1n} \\ a_{21} & a_{22} & \cdots & a_{2n} \\ \cdot & & & \cdot \\ \cdot & & & \cdot \\ \cdot & & & \cdot \\ a_{n1} & a_{n2} & \cdots & a_{nn} \end{bmatrix} \qquad (6.3\text{-}16)$$

and $\mathbf{b}$ is a column vector with constant elements b_i

$$\mathbf{b} = \begin{bmatrix} b_1 \\ b_2 \\ \cdot \\ \cdot \\ \cdot \\ b_n \end{bmatrix} \qquad (6.3\text{-}17)$$

In the bivariate case, Eq. (6.3-15) becomes

$$X_1 = a_{11} Y_1 + a_{12} Y_2 + b_1$$

$$\qquad (6.3\text{-}18)$$

$$X_2 = a_{21} Y_1 + a_{22} Y_2 + b_2$$

We shall restrict the matrix $\mathbf{A}$ to be nonsingular so that its inverse $\mathbf{A}^{-1}$ exists. In other words, no X_i is a linear combination of other $X_i{}'$ s. In this case Eq. (6.3-15) may be inverted to yield a solution for $\mathbf{Y}$. We have

$$\mathbf{A}\,\mathbf{Y} = \mathbf{X} - \mathbf{b}$$

After pre-multiplication by the inverse matrix $\mathbf{A}^{-1}$, this last expression becomes

$$\mathbf{Y} = \mathbf{A}^{-1}\,(\mathbf{X} - \mathbf{b}) \qquad (6.3\text{-}19)$$

It is clear from Eq. (6.3-15) that, for each set $\{y_i\}$ of possible values of the set of random variables $\{Y_i\}$, there exists a unique set $\{x_i\}$ of values of the set of random variables $\{X_i\}$. The converse is also true since $\mathbf{A}$ is a non-singular matrix and Eq. (6.3-19) exists. It is shown in Chapter 7 that in such cases where a one-to-one transformation exists between two sets of random variables, then the density functions are related by

$$f_{\mathbf{X}}(\mathbf{x}) = f_{\mathbf{Y}}(\mathbf{y}) \; \left| \; J\left(\frac{\mathbf{y}}{\mathbf{x}}\right) \right| \qquad (6.3\text{-}20)$$

where $J\left(\dfrac{\mathbf{y}}{\mathbf{x}}\right)$ is the Jacobian of the transformation and is given by the determinant

$$J\left(\frac{\mathbf{y}}{\mathbf{x}}\right) = J\left(\frac{y_1,\dots,y_n}{x_1,\dots,x_n}\right) = \begin{vmatrix} \dfrac{\partial y_1}{\partial x_1} & \cdots & \dfrac{\partial y_1}{\partial x_n} \\ \cdot & & \cdot \\ \cdot & & \cdot \\ \cdot & & \cdot \\ \dfrac{\partial y_n}{\partial x_1} & \cdots & \dfrac{\partial y_n}{\partial x_n} \end{vmatrix} \qquad (6.3\text{-}21)$$

This result can be justified in a somewhat heuristic fashion as follows. Consider a small volume $(dy_1 dy_2...dy_n)$ in n-space and the corresponding volume $(dx_1 dx_2...dx_n)$. As a result of the one-to-one transformation of Eq. (6.3-15), the probability that the random vector $\mathbf{Y}$ lies in $(dy_1 dy_2...dy_n)$ is the same as the probability that the random vector $\mathbf{X}$ lies in the corresponding volume $(dx_1 dx_2...dx_n)$. Hence, we have

$$f_{\mathbf{X}}(\mathbf{x}) \, dx_1 dx_2...dx_n = f_{\mathbf{Y}}(\mathbf{y}) \, dy_1 dy_2...dy_n \qquad (6.3\text{-}22)$$

This is just the coordinate transformation of Eq. (6.3-20). The absolute magnitude sign on the Jacobian arises since the probabilities involved must be non-negative.

Let us accept Eq. (6.3-20) on this basis and proceed to find the multivariate density $f_{\mathbf{X}}(\mathbf{x})$. The exponent in Eq. (6.3-13) can be written as

$$\mathbf{y}^T \mathbf{y} = [\mathbf{A}^{-1}(\mathbf{x}-\mathbf{b})]^T \ \mathbf{A}^{-1}(\mathbf{x}-\mathbf{b})$$

$$(6.3\text{-}23)$$

$$= (\mathbf{x}-\mathbf{b})^T (\mathbf{A}^{-1})^T \ \mathbf{A}^{-1} (\mathbf{x}-\mathbf{b})$$

from ordinary matrix manipulation. The mean of $\mathbf{X}$ is the column vector $\mathbf{b}$ since

$$E(\mathbf{X}) = E(\mathbf{A} \, \mathbf{Y}) + E(\mathbf{b}) = \mathbf{A} \, E(\mathbf{Y}) + \mathbf{b} = \mathbf{b} \qquad (6.3\text{-}24)$$

Recall that each component of $\mathbf{Y}$ has zero mean. The covariance matrix of $\mathbf{X}$ can be written as

$$\Lambda = E[(\mathbf{X}-\mathbf{b})(\mathbf{X}-\mathbf{b})^T] = E[(\mathbf{A} \, \mathbf{Y})(\mathbf{A} \, \mathbf{Y})^T]$$

$$(6.3\text{-}25)$$

$$= E[\mathbf{A} \, \mathbf{Y} \, \mathbf{Y}^T \, \mathbf{A}^T] = \mathbf{A} \, E[\mathbf{Y} \, \mathbf{Y}^T]\mathbf{A}^T$$

The last relation follows since $\mathbf{A}$ is a matrix of constants. The elements of column vector $\mathbf{Y}$ are independent of each other and have unit normal distributions. It should be clear that

$$E[\mathbf{Y} \, \mathbf{Y}^T] = \mathbf{I} \qquad (6.3\text{-}26)$$

where $\mathbf{I}$ is the identity matrix. Consequently, the covariance matrix becomes

$$\Lambda = \mathbf{A} \, \mathbf{A}^T \qquad (6.3\text{-}27)$$

Its inverse is

$$\Lambda^{-1} = (\mathbf{A}^{-1})^T \ \mathbf{A}^{-1}$$

as the following manipulation shows:

$$\Lambda^{-1}\Lambda = \mathbf{I} = (\mathbf{A}^{-1})^T (\mathbf{A}^{-1}\mathbf{A})\mathbf{A}^T = (\mathbf{A}^{-1})^T \ \mathbf{I} \, \mathbf{A}^T = \mathbf{A} \, \mathbf{A}^{-1} = \mathbf{I}$$

As a consequence of these relationships, Eq. (6.3-23) can be written as

$$\mathbf{y}^T \mathbf{y} = (\mathbf{x}-\mathbf{b})^T \; \Lambda^{-1} \; (\mathbf{x}-\mathbf{b}) \tag{6.3-28}$$

Let us return to Eq. (6.3-20) and find the Jacobian $J(\frac{\mathbf{y}}{\mathbf{x}})$. The determinants of Λ and Λ^{-1} are

$$| \Lambda | = | A |^2 \tag{6.3-29}$$

and

$$| \Lambda^{-1} | = | A |^{-2} \tag{6.3-30}$$

It is clear from Eqs. (6.1-14), (6.3-19), and (6.3-21) that

$$| J(\frac{\mathbf{y}}{\mathbf{x}}) | = | A |^{-1} = | \Lambda^{-1} |^{-1/2} = | \sigma^{ij} |^{1/2} \tag{6.3-31}$$

Now, if Eqs. (6.3-13), (6.3-28), and (6.3-31) are substituted into Eq. (6.3-20), the density $f_X(\mathbf{x})$ is

$$f_X(\mathbf{x}) = \frac{| \sigma^{ij} |^{1/2}}{(2\pi)^{n/2}} \; e^{-(\mathbf{x}-\mathbf{b})^T \; \Lambda^{-1}(\mathbf{x}-\mathbf{b})/2} \tag{6.3-32}$$

The result is in complete agreement with Eq. (6.3-6) for the multivariate normal distribution. The mean vector has been called $\mathbf{b}$ instead of $\mathbf{m}$.

Another result is immediately obtainable. The characteristic function of $\mathbf{X}$ can be written as

$$M_X(i\;\mathbf{u}) = E(e^{i\;\mathbf{u}^T\mathbf{X}}) = E(e^{i\;\mathbf{u}^T(\mathbf{A}\,\mathbf{Y}\,+\,\mathbf{b})})$$
$$= e^{i\;\mathbf{u}^T\,\mathbf{b}} \; M_Y(i\,\mathbf{u}^T\;\mathbf{A}) \tag{6.3-33}$$

Since $\mathbf{u}^T \mathbf{b} = \mathbf{b}^T \mathbf{u}$ and $\mathbf{u}^T \mathbf{A} = \mathbf{A}^T \mathbf{u}$, it is more convenient to write

$$M_X(i\;\mathbf{u}) = e^{i\;\mathbf{b}^T\,\mathbf{u}} \; M_Y(i\;\mathbf{A}^T\;\mathbf{u}) \tag{6.3-34}$$

Since the random vector $\mathbf{Y}$ has components which are independent and unit normal, its characteristic function is

$$M_Y(i\;\mathbf{u}) = \prod_{j=1}^{n} M_{Y_j}(i\;u_j)$$
$$= e^{-\sum_{j=1}^{n} u_j^2/2} = e^{-\mathbf{u}^T\,\mathbf{u}/2} \tag{6.3-35}$$

This relation may be used in Eq. (6.3-34) to obtain

$$M_X(i\;\mathbf{u}) = e^{i\;\mathbf{b}^T\,\mathbf{u}} \; e^{-(\mathbf{A}^T\mathbf{u})^T(\mathbf{A}^T\mathbf{u})/2} \tag{6.3-36}$$

The last exponent may be rearranged as

$$(\mathbf{A}^T\mathbf{u})^T(\mathbf{A}^T\mathbf{u}) = \mathbf{u}^T \mathbf{A}\,\mathbf{A}^T\mathbf{u} = \mathbf{u}^T \Lambda\,\mathbf{u} \tag{6.3-37}$$

Thus the characteristic function of $\mathbf{X}$ becomes

$$M_{\mathbf{X}}(i\ \mathbf{u}) = e^{i\ \mathbf{b}^T \mathbf{u}}\ e^{-\mathbf{u}^T \mathbf{\Lambda} \mathbf{u}/2} \qquad (6.3\text{-}38)$$

in exact agreement with Eq. (6.3-1) except for the substitution of $\mathbf{b}$ and $\mathbf{m}$.

Since the function of $f_{\mathbf{X}}(\mathbf{x})$ of Eq. (6.3-32) was obtained by a simple linear transformation of variables from the known density function $f_{\mathbf{Y}}(\mathbf{y})$ of Eq. (6.3-13), it is not necessary to show that $f_{\mathbf{X}}(\mathbf{x})$ is indeed a legitimate density function. It is obviously non-negative and we could, if we wished, integrate over each variable successively to show that the final result is unity. This procedure was followed in the bivariate case and is obviously very lengthy.

It was pointed out in Section 5.5 that two jointly normal random variables could be transformed linearly into two *independent* normal random variables. It should be obvious that the same procedure can be applied in the n-variate case. As a matter of fact, if we simply reverse the procedure that we have just been following, then the multivariate normal density $f_{\mathbf{X}}(\mathbf{x})$ with covariance matrix $\mathbf{\Lambda}$ and mean column vector $\mathbf{b}$ becomes the multivariate normal density $f_{\mathbf{Y}}(\mathbf{y})$ with covariance matrix $\mathbf{I}$ and mean column vector zero. In terms of matrix theory, a matrix which is positive definite may be *diagonalized* and then converted into an identity matrix.

Example 6.1

Refer to Eqs. (5.5-6) and (5.5-7). The random variables X_1 and X_2 were jointly normal with variances σ_1^2 and σ_2^2, with covariance ρ, and with zero means. Their joint density function was given by Eq. (5.5-4). The new random variables Z_1 and Z_2 were shown to be *independent* normal random variables with zero means and with variances of $\sigma_1^2(1-\rho^2)$ and σ_2^2, respectively. Let us now replace Z_1 by $\sigma_1\sqrt{1-\rho^2}\ Y_1$ and Z_2 by $\sigma_2 Y_2$. Now Y_1 and Y_2 are *independent unit normal* random variables. Also Eqs. (5.5-6) and (5.5-7) become

$$X_1 = \sigma_1\sqrt{1-\rho^2}\ Y_1 + \rho\sigma_1 Y_2$$
$$X_2 = \sigma_2 Y_2$$

In terms of the linear transformation of Eq. (6.3-15), the column vector $\mathbf{b}$ is zero and the matrix $\mathbf{A}$ becomes

$$\mathbf{A} = \begin{bmatrix} \sigma_1\sqrt{1-\rho^2} & \rho\sigma_1 \\ 0 & \sigma_2 \end{bmatrix}$$

It follows from Eq. (6.3-27) that the covariance matrix of $\mathbf{X}$ is

$$\Lambda = \mathbf{A}\,\mathbf{A}^T = \begin{bmatrix} \sigma_1^2 & \rho\sigma_1\sigma_2 \\ \rho\sigma_1\sigma_2 & \sigma_2^2 \end{bmatrix}$$

which should be compared to Eq. (6.2-1). The rest of the development proceeds as in Section 6.2. For example, the Jacobian $J(\frac{y_1, y_2}{x_1, x_2})$ becomes

$$J\left(\frac{y_1, y_2}{x_1, x_2}\right) = \begin{vmatrix} \dfrac{1}{\sigma_1\sqrt{1-\rho^2}} & -\dfrac{\rho}{\sigma_2\sqrt{1-\rho^2}} \\ 0 & \dfrac{1}{\sigma_2} \end{vmatrix} = \dfrac{1}{\sigma_1\sigma_2\sqrt{1-\rho^2}}$$

or

$$\left| J\left(\frac{y_1, y_2}{x_1, x_2}\right) \right| = |\sigma^{ij}|^{1/2}$$

Eq. (6.12-2) follows immediately except that the mean vector $\mathbf{m}$ is zero.

Note that the inverse transformation $\mathbf{Y} = \mathbf{A}^{-1}\mathbf{X}$ of this last example is given by

$$Y_1 = \frac{1}{\sigma_1\sqrt{1-\rho^2}}\, X_1 - \frac{\rho}{\sigma_2\sqrt{1-\rho^2}}\, X_2 \qquad (6.3\text{-}39)$$

$$Y_2 = \frac{1}{\sigma_2}\, X_2 \qquad (6.3\text{-}40)$$

This linear transformation $\mathbf{y} = \mathbf{A}^{-1}\mathbf{x}$ converts the positive definite quadratic form $Q(\mathbf{x}) = \mathbf{x}^T \Lambda^{-1} \mathbf{x}$ into

$$Q(\mathbf{x}) = Q(\mathbf{A}\,\mathbf{y}) = \mathbf{y}^T \mathbf{I}\,\mathbf{y} = \mathbf{y}^T\,\mathbf{y} \qquad (6.3\text{-}41)$$

The matrix Λ^{-1} has been diagonalized into the particular diagonal matrix $\mathbf{I}$. In the n-variate case, the elements of $\mathbf{A}^{-1}$ become somewhat complicated. The required $\mathbf{A}$ is the *square-root matrix* of Λ. The subject is discussed in more detail in Birnbaum (1962), Chapter 6 and in most books on matrix algebra.

6.4 *Miscellaneous Properties of the Multivariate Normal Distribution* - If the n-element column vector $\mathbf{X}$ has a multivariate normal distribution, then any subset of the elements of $\mathbf{X}$ also has a multivariate normal distribution. Perhaps the easiest way to see that this is true is to consider the multivariate normal characteristic function of Eq. (6.3-1) with exponents given explicitly by Eq. (6.3-2) and (6.3-3). If any one of the elements u_i (or any subset of these elements) is set equal to zero, the form of Eq. (6.3-1) remains unchanged. Setting one of the $u_i's$ equal to zero is equivalent to integrating the density $f_X(\mathbf{x})$ with respect to the corresponding x_i.

As in the bivariate case, the elements of $\mathbf{X}$ are statistically independent if and only if they are mutually uncorrelated. This fact is easily proved. If the X_i are mutually uncorrelated, then their covariance matrix Λ is the diagonal matrix

$$\Lambda = \begin{bmatrix} \sigma_1^2 & 0 & 0 & \cdots & 0 \\ 0 & \sigma_2^2 & 0 & \cdots & 0 \\ 0 & 0 & \sigma_3^2 & \cdots & 0 \\ \cdot & & & & \\ \cdot & & & & \\ \cdot & & & & \\ 0 & 0 & 0 & \cdots & \sigma_n^2 \end{bmatrix} \tag{6.4-1}$$

In this simple case, the inverse matrix is given by

$$\Lambda^{-1} = \begin{bmatrix} \sigma_1^{-2} & 0 & 0 & \cdots & 0 \\ 0 & \sigma_2^{-2} & 0 & \cdots & 0 \\ 0 & 0 & \sigma_3^{-2} & \cdots & 0 \\ \cdot & & & & \\ \cdot & & & & \\ \cdot & & & & \\ 0 & 0 & 0 & \cdots & \sigma_n^{-2} \end{bmatrix} \tag{6.4-2}$$

and, in addition,

$$|\Lambda^{-1}|^{1/2} = |\sigma^{ij}|^{1/2} = \prod_{j=1}^{n} \frac{1}{\sigma_j} \tag{6.4-3}$$

The joint density $f_X(\mathbf{x})$ of Eq. (6.3-6) now becomes

$$f_X(\mathbf{x}) = \prod_{j=1}^{n} \frac{1}{\sqrt{2\pi\sigma_j^2}} \, e^{-(x_j - m_j)^2/2\sigma_j^2} \tag{6.4-4}$$

or

$$f_X(\mathbf{x}) = \prod_{j=1}^{n} f_{X_j}(x_j) \tag{6.4-5}$$

where $f_{X_j}(x_j)$ is a normal density function with mean m_j and variance σ_j^2. Thus the joint density function factors into the product of the univariate densities and the set of X_i are independent normal random variables.

Various higher order moments are relatively easy to obtain from the moment generating function of Eq. (6.3-4) by direct differentiation. The basic definition of the moment generating function is

$$M_X(t) = E(e^{t^T X}) = E\left[exp\left(\sum_{j=1}^{n} t_i X_i\right)\right] \qquad (6.4-6)$$

It is easy to see that differentiation yields

$$E(X_i^a X_j^b X_k^c \ldots) = \frac{\partial^{a+b+c+\ldots}}{\partial t_i^a \partial t_j^b \partial t_k^c \cdots} M_X(t) \Big|_{\substack{t_m = 0 \\ \text{all } m}} \qquad (6.4-7)$$

where $a, b, c, \cdots$ are integers. For the multivariate normal distribution, the differentiation is straightforward. Some useful results are:

in the univariate case

$$\frac{\partial}{\partial t_1} M_{x_1}(t_1)\Big|_{t_1=0} = E(X_1) = m_1 \qquad (6.4-8)$$

in the bivariate case

$$\frac{\partial^2}{\partial t_1 \partial t_2} M_{X_1, X_2}(t_1, t_2)\Big|_{t_1=t_2=0} = E(X_1 X_2) \qquad (6.4-9)$$

in the 3-variate case

$$\frac{\partial^3}{\partial t_1 \partial t_2 \partial t_3} M_{X_1, X_2, X_3}(t_1, t_2, t_3)\Big|_{t_1=t_2=t_3=0} = E(X_1 X_2 X_3)$$

$$(6.4-10)$$

$$= m_1 E(X_2 X_3) + m_2 E(X_1 X_3) + m_3 E(X_1 X_2) - 2m_1 m_2 m_3$$

in the 4-variate case

$$\frac{\partial^4}{\partial t_1 \partial t_2 \partial t_3 \partial t_4} M_{X_1, X_2, X_3, X_4}(t_1, t_2, t_3, t_4)\Big|_{t_1=t_2=t_3=t_4=0}$$

$$= E(X_1 X_2 X_3 X_4)$$

$$(6.4-11)$$

$$= E(X_1 X_2)E(X_3 X_4) + E(X_1 X_3)E(X_2 X_4)$$

$$+ E(X_1 X_4)E(X_2 X_3) - 2m_1 m_2 m_3 m_4$$

It is clear that higher moments, such as the fourth moment $E(X_1 X_2 X_3 X_4)$, are all expressible in terms of first and second moments. This is to be expected since the multivariate density function of Eq. (6.3-6) depends only on means and covariances. Higher order moments such as Eq. (6.4-11) arise in problems involving the non-linear transformation of normal random variables.

6.5 *Linear Transformations on Normal Random Variables* - It has already been shown that sums of normal random variables are normal and that linear transformations on certain sets of jointly normal random variables give other sets of jointly normal random variables. The discussion of Section 6.3 may be modified slightly to show that linear transformations on arbitrary jointly normal random variables yield jointly normal random variables. This is one of the most useful properties of normal random variables.

Consider the n-variate normal random column vector $\mathbf{X}$ with characteristic function $M_{\mathbf{X}}(i\ \mathbf{u})$ given by Eq. (6.3-1). Let an arbitrary linear transformation on $\mathbf{X}$ be given by

$$\mathbf{Z} = \mathbf{B}\,\mathbf{X} \tag{6.5-1}$$

where $\mathbf{Z}$ is an n-variate random column vector and $\mathbf{B}$ is an $n \times n$ constant matrix with elements b_{ij}. For convenience, we have dropped the added constant column vector such as occurred in Eq. (6.3-15). The means of $\mathbf{Z}$ are given by the column vector

$$E\,(\mathbf{Z}) = \mathbf{m}_{\mathbf{Z}} = \mathbf{B}\,\mathbf{m} \tag{6.5-2}$$

where the elements of $\mathbf{m}$ are the means of the X_i. The covariance matrix of $\mathbf{Z}$ is

$$
\begin{aligned}
\mathbf{\Lambda}_{\mathbf{Z}} &= E\,[(\mathbf{Z}-\mathbf{m}_{\mathbf{Z}})(\mathbf{Z}-\mathbf{m}_{\mathbf{Z}})^{T}\,] \\
&= \mathbf{B}\,E\,[(\mathbf{X}-\mathbf{m})(\mathbf{X}-\mathbf{m})^{T}\,]\mathbf{B}^{T} \\
&= \mathbf{B}\,\mathbf{\Lambda}\,\mathbf{B}^{T}
\end{aligned} \tag{6.5-3}
$$

where $\mathbf{\Lambda}$ is the covariance matrix of $\mathbf{X}$.

Consider now the characteristic function $M_{\mathbf{X}}(i\ \mathbf{u})$ of Eq. (6.3-1) and make the substitution

$$\mathbf{u} = \mathbf{B}^{T}\,\mathbf{w} \tag{6.5-4}$$

for the dummy variable. Here $\mathbf{w}$ is the real column vector

$$\mathbf{w} = \begin{bmatrix} w_1 \\ w_2 \\ \cdot \\ \cdot \\ \cdot \\ w_n \end{bmatrix} \tag{6.5-5}$$

Now $M_{\mathbf{X}}(\mathbf{x})$ becomes

$$M_{\mathbf{X}}(i\ \mathbf{u}) = \exp[i\,(\mathbf{B}\,\mathbf{m})^{T}\,\mathbf{w} - \mathbf{w}^{T}\,(\mathbf{B}\,\mathbf{\Lambda}\,\mathbf{B}^{T}\,)\mathbf{w}/2] \tag{6.5-6}$$

With the aid of Eqs. (6.5-2) and (6.5-3), this last expression may be written as

$$M_{\mathbf{X}}(i \ \mathbf{u}) = \exp[i \ \mathbf{m}_Z^T \ \mathbf{w} - \mathbf{w}^T \ \Lambda_Z \ \mathbf{w}/2] \qquad (6.5\text{-}7)$$

This is the characteristic function of an n-variate normal random vector with mean vector $\mathbf{m}_Z$, covariance matrix Λ_Z, and dummy variable $\mathbf{w}$. Note that the characteristic function of $\mathbf{Z}$ is defined as

$$M_Z(i \ \mathbf{w}) = E(e^{i \ \mathbf{w}^T \ \mathbf{Z}}) \qquad (6.5\text{-}8)$$

Since $\mathbf{Z} = \mathbf{B} \ \mathbf{X}$, this expression becomes

$$M_Z(i \ \mathbf{w}) = E(e^{i \ \mathbf{w}^T \ \mathbf{B} \ \mathbf{X}}) = E(e^{i \ \mathbf{u}^T \ \mathbf{X}}) = M_{\mathbf{X}}(i \ \mathbf{u}) \qquad (6.5\text{-}9)$$

A comparison of Eqs. (6.5-7) and (6.5-9) shows that $\mathbf{Z}$ is an n-variate normal random vector with mean vector $\mathbf{m}_Z = \mathbf{B} \ \mathbf{m}$ and covariance matrix $\Lambda_Z = \mathbf{B} \ \Lambda \ \mathbf{B}^T$. Hence arbitrary linear transformations on jointly normal random variables yield jointly normal random variables.

PROBLEMS

1. Let $\mathbf{X}$ be the n-variate normal random vector. Show that the conditional distribution of any subset of $n-1$ components of $\mathbf{X}$ given the remaining component is an (n-1)-variate normal random vector.

2. The 3-variate normal random vector $\mathbf{X}$ has the density function

    $$f_{\mathbf{X}}(\mathbf{x}) = C \ e^{-(2x_1^2 - x_1 x_2 + x_2^2 - 2x_1 x_3 + 4x_3^2)/2}$$

    a) Find the constant C.
    b) Find the covariance matrix Λ for $\mathbf{X}$.
    c) Prove that $f_{\mathbf{X}}(\mathbf{x})$ is a legitimate multivariate normal density, that is, show that Λ is positive definite.

3. Derive Eqs. (6.4-10) and (6.4-11).

Chapter 7

THE TRANSFORMATION OF RANDOM VARIABLES

In this chapter we consider the problems that arise when a random variable X is transformed to a new random variable Y through some functional relationship $y = h(x)$. We shall restrict our attention to cases where y is a single-valued function of x. The usual problem is, given the density function of X, to find the density function of $Y = h(X)$. In a more general setting, we may have a set of random variables $X_1, X_2, ..., X_n$ related to another set $Y_1, Y_2, ..., Y_m$ through the m known single-valued functions.

$$y_i = h(x_1, x_2, ..., x_n) \quad , \quad i = 1, 2, ..., m$$

Some simple problems of this type have already been encountered. In particular the case where a random variable Y is a linear function of a random variable X or of a set $X_1, X_2, ..., X_n$ has been treated in some special cases. One of the most interesting results that has been obtained thus far is that a linear combination of jointly normal random variables is a normal random variable.

In the discussion thus far the function $y = h(x)$ has been restricted only to be single-valued. If X is a discrete random variable, then further restrictions are unnecessary and $h(x)$ may be arbitrary. If X is a continuous random variable, then $h(\bullet)$ must be what is called a *Borel function*. Supposed that $h(\bullet)$ maps points on a real line R into points on another real line S. The $h(\bullet)$ is a Borel function iff for every Borel set* B in R, the set $C = h^{-1}(B)$ is a Borel set in S. Recall that the random variable X was defined originally in Chapter 2 as a mapping from an abstract probability space to the real line in such a way that events in the original space could be well defined as sets of points on the real line R. The restriction of $h(\bullet)$ to be a Borel function insures that events in the original space continue to correspond to well-defined sets of points on the real line S. As a practical matter, the idea of Borel function is so general that no $h(\bullet)$ which is not a Borel function will ever be encountered in applications. A further discussion is beyond the scope of this treatment. Readers who are further interested might begin by reading Feller (1966), Chapter IV.

---

* Borel sets have been mentioned only slightly in previous parts of this book. The Borel sets on the real line are the completely additive class of sets generated by the intervals on the real line. Our original definition of a random variable was such as to insure that every Borel set on the real line corresponded to an event in the abstract probability space.

7.1 *Discrete Random Variables* - As pointed out in Eq. (2.2-2), a discrete random variable X can be written as

$$X = \sum_i x_i \, I_{[X=x_i]} \qquad (7.1\text{-}1)$$

where the indicator $I_{[X=x_i]}$ is defined by

$$I_{[X=x_i]} = \begin{cases} 1 & , \ X=x_i \\ 0 & , \ X \neq x_i \end{cases} \qquad (7.1\text{-}2)$$

If a *one-to-one transformation* $y = h(x)$ exists, it is clear that the random variable $Y = h(X)$ is given by

$$Y = \sum_i h(x_i) \, I_{[X=x_i]} \qquad (7.1\text{-}3)$$

If the density function $f_X(x_i) = P(X=x_i)$ is known, it is apparent that the density function of Y is just

$$f_Y(y) = P(Y=y) = \sum_{\{x_i \mid h(x_i)=y\}} f_X(x_i) \qquad (7.1\text{-}4)$$

The problem has been reduced to a simple change of variable.

Example 7.1

Let X have a density function given by

$$f_X(x) = \begin{cases} 1/4 & , \ x=1,2,3,4 \\ 0 & , \ \text{otherwise} \end{cases}$$

Consider the transformation $y=2x$. It is clear that the density function of $Y=2X$ is

$$f_Y(y) = \begin{cases} 1/4 & , \ y=2,4,6,8 \\ 0 & , \ \text{otherwise} \end{cases}$$

If the transformation $y=h(x)$ is not one-to-one, then several values of x may correspond to a single value of y and each of these contributions must be counted in arriving at the probability $P(Y=y_i)$. In such cases the problem may be resolved by straightforward counting as the following example shows.

Example 7.2

Let X have the same density function as in Example 7.1 but consider the transformation

$$y = (x-2)(x-3)$$

Since X can take on only four values, a table is easily constructed:

x	1	2	3	4
y	2	0	0	2

It is obvious that the density function of Y is

$$f_Y(y) = \begin{cases} 1/2 & , \ y=0 \\ 1/2 & , \ y=2 \\ 0 & , \ \text{otherwise} \end{cases}$$

Thus the transformation of variables in the discrete case is straightforward, although it is conceivable that it might be necessary to construct a large table of corresponding value of x and y. The continuous case will turn out to be more complex.

7.2 *Continuous Random Variables; the Univariate Case* - We begin with the idea of a *monotonic function*. A function $h(x)$ is *monotonically increasing* if $h(x_2) > h(x_1)$ when $x_2 > x_1$. Similarly, a function $h(x)$ is *monotonically decreasing* if $h(x_2) < h(x_1)$ when $x_2 > x_1$. It is apparent that, for monotonic functions, there is a one-to-one correspondence between $h(x)$ and x; that is, for each y in the range of h there is one and only one value of x such that $h(x)=y$.

Consider first the case where $y=h(x)$ is monotonically increasing and possesses a derivative everywhere. Let the random variable X have a density function $f_X(x)$ and a distribution function $F_X(x)$. It is desired to find the density function $f_Y(y)$ and distribution function $F_Y(y)$ of the random variable $Y=h(X)$. Choose some particular value of x and the corresponding value of y, say x_1 and $y_1 = h(x_1)$. It is clear from Fig. 7.1 that

$$P(Y \leq y_1) = P(X \leq x_1) \tag{7.2-1}$$

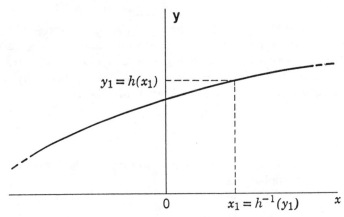

Fig. 7.1 - A monotonically increasing function

since $h(x)$ is monotonically increasing. In other words, for arbitrary x, the relationship holds that

$$F_Y(y) = F_X(x) \qquad (7.2\text{-}2)$$

where y and x are related through

$$y = h(x) \qquad (7.2\text{-}3)$$

Since x and y are related one-to-one, a unique inverse h^{-1} exists for a function h, that is,

$$x = h^{-1}(y) \qquad (7.2\text{-}4)$$

where the notation h^{-1} means

$$h(x) = h[h^{-1}(y)] = y \qquad (7.2\text{-}5)$$

If X is a continuous random variable, then Y is also and both sides of Eq. (7.2-2) may be differentiated with respect to y to yield

$$\frac{dF_X(y)}{dy} = f_Y(y) = \frac{dF_X(x)}{dx}\frac{dx}{dy} = f_X(x)\frac{dx}{dy} \qquad (7.2\text{-}6)$$

where the derivative

$$\frac{dx}{dy} = \frac{d[h^{-1}(y)]}{dy}$$

is always positive since $h(x)$ is a monotonically increasing function. Thus the density functions of Y and X obey the relationship

$$f_Y(y) = f_X(x)\frac{dx}{dy} \qquad (7.2\text{-}7)$$

where y and x are related by Eqs. (7.2-3) and (7.2-4), which are equivalent. To be formally correct, this last equation should really be written as

$$f_Y(y) = f_X[h^{-1}(y)]\frac{d[h^{-1}(y)]}{dy} \qquad (7.2\text{-}8)$$

but this will usually not be done.*

Example 7.3

Let $y = ax$, where $a > 0$, and let X be a random variable with the unit normal density function

$$f_X(x) = \frac{1}{\sqrt{2\pi}}\, e^{-x^2/2} \quad , \quad -\infty < x < \infty$$

---

*Note that h^{-1} is defined only on the range of h, and the density f_Y will be zero for all y not in the range of h.

It is immediately clear that the density function of Y is simply

$$f_Y(y) = f_X\left(\frac{y}{a}\right)\frac{1}{a} = \frac{1}{\sqrt{2\pi}\,a}\,e^{-y^2/2a^2}\,, \quad -\infty < y < \infty$$

As expected, Y is normal with mean zero and variance a^2.

Note that the problem could have been solved as easily in this case from Eq. (7.2-2). We write

$$P\{Y \le y\} = P\{X \le x\} = P\{X \le y/a\}$$

or

$$P\{Y \le y\} = F_Y(y) = \frac{1}{\sqrt{2\pi}}\int_{-\infty}^{y/a} e^{-w^2/2}\,dw$$

The linear change in variable $y = aw$ in the definite integral gives

$$P\{Y \le y\} = \frac{1}{\sqrt{2\pi}\,a}\int_{-\infty}^{y} e^{-y^2/2a^2}\,dy$$

This is the distribution function of a normal random variable Y with mean zero and variance a^2. Differentiation of both sides by y yields the density function $f_Y(y)$ obtained previously.

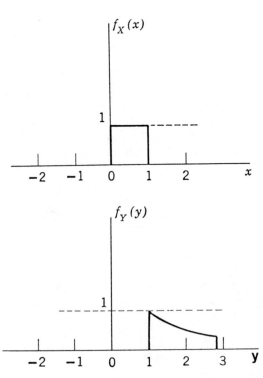

Fig. 7.2 - A plot of the densities $f_X(x)$ and $f_Y(y)$ of Example 7.4

Example 7.4

Let X be uniformly distributed in [0,1] and let the transformation be $y = e^x$. Then, from Eq. (7.2-7), the density of y is

$$f_Y(y) = f_X(x) e^{-x} = 1/y \quad , \quad 0 < x < 1$$

For $0 < x < 1$, it is clear that $1 < y < 2.718...$ or

$$f_Y(y) = \begin{cases} 1/y & , \quad 1 \le y \le 2.718... \\ 0 & , \quad \text{otherwise} \end{cases}$$

The function $f_Y(y)$ is a density function since it is non-negative and

$$\int_1^e (1/y) \, dy = \log_e y \Big|_1^e = 1$$

The two densities $f_X(x)$ and $f_Y(y)$ are plotted in Fig. 7.2.

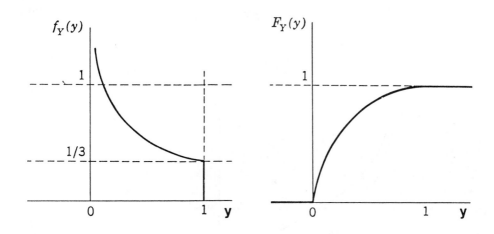

Fig. 7.3 - Distribution function and density function
for Example 7.5

Example 7.5

As in Example 7.4, let X be uniformly distributed in [0,1] and let $y = x^3$. In this case, let us find the density function $f_Y(y)$ by first finding the distribution function $F_Y(y)$. We have

$$F_Y(y) = P\{Y \leq y\} = P\{X^3 \leq y\} = P\{X \leq y^{1/3}\}$$

It follows immediately that

$$F_Y(y) = P\{X \leq y^{1/3}\} = \begin{cases} 0 & , \ y < 0 \\ y^{1/3}, & 0 \leq y \leq 1 \\ 1 & , \ y > 1 \end{cases}$$

On differentiating, we obtain the density

$$f_Y(y) = \frac{dF_Y(y)}{dy} = \begin{cases} 0 & , \ y < 0 \\ (1/3)y^{-2/3}, & 0 \leq y \leq 1 \\ 0 & , \ y > 1 \end{cases}$$

Both $F_Y(y)$ and $f_Y(y)$ are plotted in Fig. 7.3. It is left to the reader to show that $F_Y(y)$ is a distribution function, $f_Y(y)$ is a density function, and that the same results would be obtained using Eq. (7.2-7)

Consider next the case where $y = h(x)$ is a monotonically decreasing function. As shown in Fig. 7.4 it is clear that

$$P\{Y \leq y\} = P\{X \geq x\} \tag{7.2-9}$$

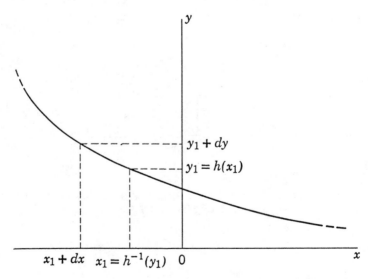

Fig. 7.4 - A monotonically decreasing function

Note that the inequality on the right side of this expression is reversed. Equation (7.2-9) is the same as

$$F_Y(y) = 1 - F_X(x) \qquad (7.2\text{-}10)$$

keeping in mind that $y = h(x)$ and that $h^{-1}(y) = x$. Now differentiation of both sides of Eq. (7.2-10) yields

$$f_Y(y) = -f_X(x)\,\frac{dx}{dy} \qquad (7.2\text{-}11)$$

However, the derivative

$$\frac{dx}{dy} = \frac{d\,[h^{-1}(y)]}{dy}$$

is always negative since $h(x)$ is a monotonically decreasing function. Thus both Eqs. (7.2-7) and (7.2-11) can be written as

$$f_Y(y) = f_X(x)\,\Big|\,\frac{dx}{dy}\,\Big| \qquad (7.2\text{-}12)$$

where $\big|\,\frac{dx}{dy}\,\big|$ indicates the magnitude of the derivative and where $y = h(x)$ is constrained to be a one-to-one continuously differentiable transformation with inverse $x = h^{-1}(y)$. Again, Eq. (7.2-12) holds only on the range of h.

Example 7.6

Let X be uniformly distributed in $[0,1]$ and let $y = e^{-x}$. It follows immediately form Eq. (7.2-12) that

$$f_Y(y) = f_X(x)\,e^x = 1/y$$

or

$$f_Y(y) = \begin{cases} 1/y, & (1/e) \le y \le 1 \\ 0, & \text{otherwise} \end{cases}$$

It may be interesting to the reader to plot $f_Y(y)$ to compare to Example 7.4 and Fig. 7.2.

Another way to look at the problem when one-to-one transformations $y = h(x)$ are involved is as follows. Consider Fig. 7.4 where a differential interval dy and the corresponding interval dx are shown. It does not matter whether y is monotonically increasing or decreasing. The probability that the random variable Y lies in some differential interval dy is equal to the probability that X lies in the corresponding interval dx. In other words,

$$f_Y(y)\,|\,dy\,| = f_X(x)\,|\,dx\,| \qquad (7.2\text{-}13)$$

where the absolute magnitude signs are necessary since both sides of this expression must be non-negative. On rearranging, this expression becomes Eq. (7.2-12). Note that Eq. (7.2-13) could also be written as

$$f_Y(y) = f_X(x) \left| J\left(\frac{x}{y}\right) \right| \qquad (7.2\text{-}14)$$

where $J\left(\frac{x}{y}\right)$ is the Jacobian of the transformation $y = h(x)$ and is

$$J\left(\frac{x}{y}\right) = \frac{dx}{dy} \qquad (7.2\text{-}15)$$

It might be useful to point out that when $y = h(x)$ is a one-to-one continuously differentiable transformation, then $J(x/y)$ is either positive or negative, but does not change sign or become zero. Actually we are only interested in that interval of the x-line where $f_X(x)$ is non-zero. In that interval, the Jacobian must retain the same sign if Eq. (7.2-14) is to hold.

Consider now the more general problem where $y = h(x)$ is not a monotone function of x. In this case, several values of $y = h(x)$ may correspond to a single value of x; in other words the inverse function h^{-1} is multivalued. It is true in general that the distribution function of the random variable Y is determined from

$$P\{Y \le y\} = P\{X = \text{those } x \text{ for which } Y \le y\} \qquad (7.2\text{-}16)$$

Thus the distribution function of Y becomes

$$F_Y(y) = P\{Y \le y\} = \int_R f_X(x)\, dx \qquad (7.2\text{-}17)$$

where $f_X(x)$ is the density function of X and the region R is that interval or set of intervals defined by

$$R = \{x \mid h(x) \le y\} \qquad (7.2\text{-}18)$$

The density function of Y can then be obtained by differentiation. For simple transformations, the procedure is straightforward as several examples will illustrate.

Example 7.7 - The Square-Law Transformation

Let $y = x^2$ as illustrated in Fig. 7.5. It is clear that, for each value of y, there corresponds two values of x. We have

$$P\{Y \le y\} = P\{-\sqrt{y} \le X \le \sqrt{y}\}$$

or

$$F_Y(y) = \int_{-\sqrt{y}}^{\sqrt{y}} f_X(x)\, dx$$

The density function of Y can now be found by differentiation. As pointed out in Appendix E, the derivative is

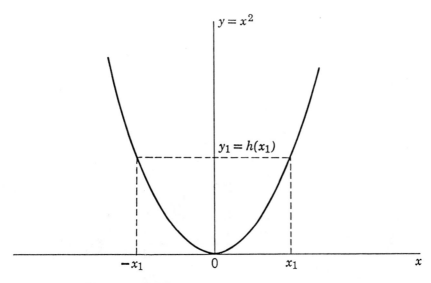

Fig. 7.5 -The square-law transformation

$$f_Y(y) = \frac{dF_Y(y)}{dy} = f_X(\sqrt{y}\,) \frac{d}{dy} \sqrt{y} - f_X(-\sqrt{y}\,) \frac{d}{dy} (-\sqrt{y}\,)$$

or

$$\dot{f}_Y(y) = \begin{cases} 0 & , \ y < 0 \\ \dfrac{f_X(\sqrt{y}\,)+f_X(-\sqrt{y}\,)}{2\sqrt{y}} & , \ y \geq 0 \end{cases}$$

As a specific example, let X be unit normal so that

$$f_X(x) = \frac{1}{\sqrt{2\pi}}\, e^{-x^2/2} \quad , \quad -\infty < x < \infty$$

Then the density of Y is

$$f_Y(y) = \begin{cases} 0 & , \ y < 0 \\ \dfrac{1}{\sqrt{2\pi}}\, y^{-1/2}\, e^{-y/2} & , \ y \geq 0 \end{cases}$$

Note that $f_Y(y)$ becomes infinite as y approaches zero from the right, but the area under the curve of $f_Y(y)$ remains unity:

$$\lim_{\substack{h \to 0 \\ h > 0}} \int_h^\infty f_Y(y) = 1$$

The random variable Y is *chi-squared distributed* with one degree of freedom [Kendall and Stuart, 1977]. Both the unit normal and the chi-squared densities are shown in Fig. 7.6.

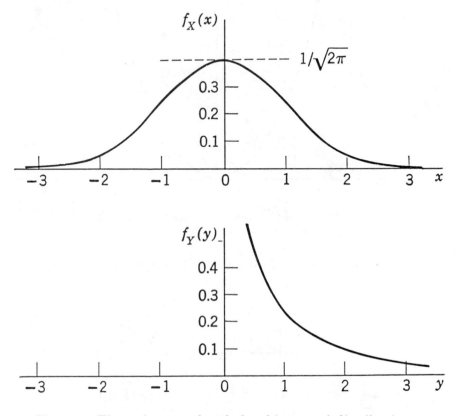

Fig. 7.6 - The unit normal and the chi-squared distribution

Example 7.8 - The Absolute-Value Transformation

Let $y = |x|$ as shown in Fig. 7.7. Again, for each value of y, there corresponds two values of x. As before

$$F_Y(y) = \int_{-y}^{y} f_X(x)\, dx$$

or

$$f_Y(y) = \begin{cases} 0 & y < 0 \\ f_X(y) + f_X(-y) & y \geq 0 \end{cases}$$

Let us look somewhat more closely at the absolute value transformation of Fig. 7.7. For a given value of y, there corresponds two values

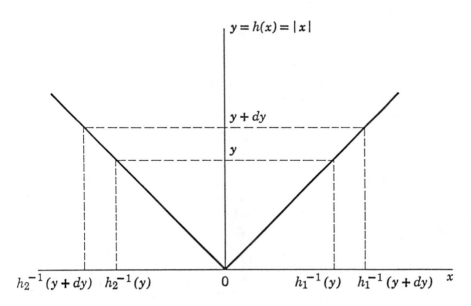

Fig. 7.7 - The absolute-value transformation

of x which, for convenience, we call $h_1^{-1}(y)$ and $h_2^{-1}(y)$ as shown in Fig. 7.7. In this case $h_1^{-1}(y) = x$ and $h_2^{-1}(y) = -x$. As pointed out previously the probability that the random variable Y lies in some small interval $(y, y + dy)$ is given by the Mean-Value Theorem as

$$P\{y \leq Y \leq y + dy\} \approx f_Y(y)\, dy \qquad (7.2\text{-}19)$$

This expression is accurate to second order terms in dy. This probability is also

$$f_Y(y)\,dy \approx P\{h_1^{-1}(y) \leq X \leq h_1^{-1}(y + dy)\}$$
$$+ P\{h_2^{-1}(y) \leq X \leq h_2^{-1}(y + dy)\} \qquad (7.2\text{-}20)$$

or, in terms of the density function $f_X(x)$,

$$f_Y(y)\,dy \approx f_X[h_1^{-1}(y)][h_1^{-1}(y + dy) - h_1^{-1}(y)]$$
$$+ f_X[h_2^{-1}(y)][h_2^{-1}(y) - h_2^{-1}(y + dy)] \qquad (7.2\text{-}21)$$

On dividing through by dy and going to the limit, we obtain

$$f_Y(y) = f_X[h_1^{-1}(y)]\left| \frac{dh_1^{-1}(y)}{dy} \right|$$

$$+ f_X[h_2^{-1}(y)]\left| \frac{dh_2^{-1}(y)}{dy} \right| \qquad (7.2\text{-}22)$$

It is clear that the same result could have been obtained by considering separately the region $x>0$ where $y=h(x)$ is monotone increasing and the region $x<0$ where $y=h(x)$ is monotone decreasing. The two contributions of x to a given value of y could then have been added.

This procedure is obviously capable of generalization in a straightforward fashion to the case where n values of x correspond to each value of y so long as $y=h(x)$ is either monotone increasing or monotone decreasing in each region. It should also be somewhat obvious that the procedure does not differ essentially from writing the distribution function $F_Y(y)$ according to Eq. (7.2-17) and then differentiating.

Thus Eq. (7.2-22) may be generalized in the following way. The transformation is $y=h(x)$. Let the domain of h be subdivided into a set of disjoint intervals I_j that are exhaustive. Thus the domain of h can be expressed as a union of the intervals $I_1, I_2, ..., I_j, ...$ Form the I_j such that h is monotone increasing or monotone decreasing in each I_j. Assume that h is differentiable in each I_j. The inverse of h in I_j will be denoted by h_j^{-1}. Then the density function of Y is given by the sum

$$f_Y(y) = \sum_j f_X[h_j^{-1}(y)] \left| \frac{dh_j^{-1}(y)}{dy} \right| \qquad (7.2\text{-}23)$$

It is assumed, of course, that the summand

$$f_X[h_j^{-1}(y)] \left| \frac{dh_j^{-1}(y)}{dy} \right|$$

is zero if y does not belong to the domain of h_j^{-1}.

Example 7.9

Let $y=\cos X$ where X is a random variable uniformly distributed in the interval $[-\pi,\pi]$; that is;

$$f_X(x) = \begin{cases} 1/2\pi & , \ -\pi \leq x \leq \pi \\ 0 & , \ \text{elsewhere} \end{cases}$$

It is clear that in the interval $(-\pi,0)$, the function $y=\cos x$ is monotone increasing and that in the interval $(0,\pi)$ it is monotone decreasing. Furthermore, we have

$$x = \arccos y, \quad \frac{dx}{dy} = \frac{-1}{\sqrt{1-y^2}}$$

Thus $f_Y(y)$ is non-zero only in the interval $[-1,1]$ and Eq. (7.2-23) becomes

$$f_Y(y) = \frac{1}{2\pi} \frac{1}{\sqrt{1-y^2}} + \frac{1}{2\pi} \frac{1}{\sqrt{1-y^2}}$$

or

$$f_Y(y) = \begin{cases} \dfrac{1}{\pi} \dfrac{1}{\sqrt{1-y^2}} & , \ -1 \le y \le 1 \\ 0 & , \ \text{elsewhere} \end{cases}$$

This result was obtained in a way that was not essentially different from that used for the square-law device $y = x^2$ of Example 7.7. For each value of y, there corresponded two values of x of the same magnitude but opposite sign. The two densities $f_X(x)$ and $f_Y(y)$ are plotted in Fig. 7.8.

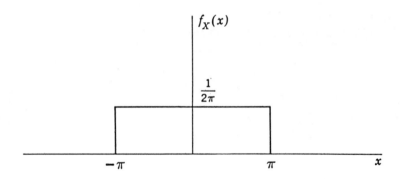

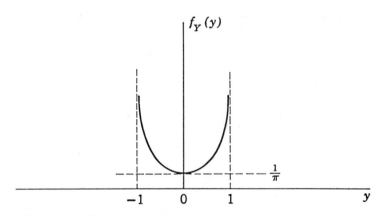

Fig. 7.8 - The transformation $y = \cos X$ when X has a symmetrical uniform distribution

Example 7.10

The previous example was simple because of the symmetry of the transformation. Suppose we assume that $f_X(x)$ is uniformly distributed in the interval $[-\pi/2,\pi]$ as shown in Fig. 7.9(a). Now $f_X(x)$ is monotone increasing in $(-\pi/2,0)$ and monotone decreasing in $(0,\pi)$ as before. In the y-interval $(-1,0)$ there is only one contribution from $f_X(x)$ while in the y-interval $(0,1)$ there are two as before. Thus $f_Y(y)$ becomes

$$f_Y(y) = \begin{cases} \dfrac{2}{3\pi}\,\dfrac{1}{\sqrt{1-y^2}} & , \ -1 \le y < 0 \\[3ex] \dfrac{4}{3\pi}\,\dfrac{1}{\sqrt{1-y^2}} & , \ 0 < y \le 1 \\[3ex] 0 & , \ \text{elsewhere} \end{cases}$$

As before, $f_X(x)$ and $f_Y(y)$ are plotted in Fig. 7.9. Note the lack of symmetry in $f_Y(y)$ and the jump at $y=0$.

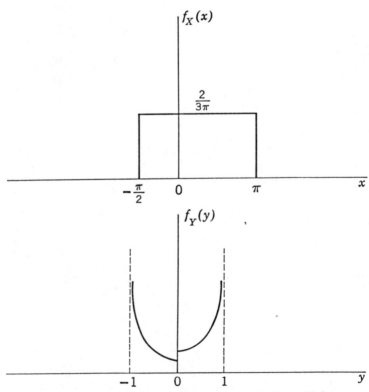

Fig. 7.9 - The transformation $Y=\cos X$ when X has an asymmetrical uniform distribution

7.3 *Continuous Random Variables; the Bivariate Case* - The problem of functional transformations is considerably more complicated for bivariate distributions although the principles remain the same as in the univariate case. Let us begin by restricting attention to the transformations

$$y_1 = h_1(x_1, x_2) \tag{7.3-1}$$

and

$$y_2 = h_2(x_1, x_2) \tag{7.3-2}$$

which are assumed to form a one-to-one continuously differentiable transformation of the random variables X_1 and X_2 into Y_1 and Y_2. By continuously differentiable we mean that the first partial derivatives $\dfrac{\partial y_j}{\partial x_i}$ exist and are continuous. On this assumption, the Jacobian of the transformation exists, is continuous, and will be assumed to be non-zero. The Jacobian is given by the determinant

$$J\left(\frac{y_1, y_2}{x_1, x_2}\right) = \begin{vmatrix} \dfrac{\partial y_1}{\partial x_1} & \dfrac{\partial y_1}{\partial x_2} \\ \dfrac{\partial y_2}{\partial x_1} & \dfrac{\partial y_2}{\partial x_2} \end{vmatrix} \neq 0 \tag{7.3-3}$$

In this case an inverse transformation

$$x_1 = g_1(y_1, y_2) \tag{7.3-4}$$

$$x_2 = g_2(y_1, y_2) \tag{7.3-5}$$

exists which is also one-to-one and the Jacobian of this transformation exists also, is continuous, and is given by

$$J\left(\frac{x_1, x_2}{y_1, y_2}\right) = \left[J\left(\frac{y_1, y_2}{x_1, x_2}\right) \right]^{-1} \tag{7.3-6}$$

As discussed previously, the probability that X_1 and X_2 lie in some differential area $dx_1 dx_2$ is equal to the probability that Y_1 and Y_2 lie in the corresponding differential area $dy_1 dy_2$. This fact may be written as

$$f_{X_1, X_2}(x_1, x_2) dx_1 dx_2 = \pm f_{Y_1, Y_2}(y_1, y_2) dy_1 dy_2 \tag{7.3-7}$$

where f_{X_1, X_2} is the joint density function of X_1 and X_2 and f_{Y_1, Y_2} is the joint density function of Y_1 and Y_2. The $\pm$ signs arise because the differential areas may be either of the same or opposite signs while both sides of Eq. (7.3-7) must have the same sign. Equation (7.3-7) may be rewritten as

$$f_{X_1,X_2}(x_1,x_2) = f_{Y_1,Y_2}(y_1,y_2) \left| J\left(\frac{y_1,y_2}{x_1,x_2} \right) \right| \qquad (7.3\text{-}8)$$

or as

$$f_{Y_1,Y_2}(y_1,y_2) = f_{X_1,X_2}(x_1,x_2) \left| J\left(\frac{x_1,x_2}{y_1,y_2} \right) \right| \qquad (7.3\text{-}9)$$

from Eq. (7.3-6). Note the absolute magnitude signs on the Jacobian J; these follow from Eq. (7.3-7) and from the fact that probabilities must be non-negative. Recall how they arose in the univariate case from differentiating the distribution function. If the same procedure were followed here, a minus sign would arise in the differentiation whenever the Jacobian was negative as in Eq. (7.2-11).

Example 7.11

Let $Y_1 = X_1 + X_2$ where X_1 and X_2 are independent random variables with density functions $f_{X_1}(x_1)$ and $f_{X_2}(x_2)$ so that $f_{X_1,X_2}(x_1,x_2) = f_{X_1}(x_1)f_{X_2}(x_2)$. The problem is to find the density function of Y_1.

For convenience, let $Y_2 = X_2$. The Jacobian is

$$J\left(\frac{y_1,y_2}{x_1,x_2} \right) = \begin{vmatrix} 1 & 1 \\ 0 & 1 \end{vmatrix} = 1$$

It follows from Eq. (7.3-8) that the joint density of Y_1 and Y_2 is

$$f_{Y_1,Y_2}(y_1,y_2) = f_{X_1}(y_1-y_2)f_{X_2}(y_2)$$

The univariate density $F_{Y_1}(y_1)$ may be found by integrating with respect to y_2 or

$$f_{Y_1}(y_1) = \int_S f_{X_1}(y_1-y_2)f_{X_2}(y_2)dy_2$$

The interval S must be chosen by considering the region where f_{Y_1,Y_2} is non-zero. Later examples will indicate that the choice of S may require some thought. This last expression is the *convolution* of f_{X_1} and f_{X_2} and has been encountered before in Section 3.7. It is sometimes written as $f_{X_1}*f_{X_2}(y_1)$.

Example 7.12

Let Y_1 and Y_2 be general linear transformations of X_1 and X_2, that is

$$Y_1 = b_{11} X_1 + b_{12} X_2 + c_1$$

$$Y_2 = b_{21} X_2 + b_{22} X_2 + c_2$$

where b_{ij} and c_i are constants. As in Eq. (6.3-15), these last two expressions may be written in matrix form as

$$\mathbf{Y} = \mathbf{B} \mathbf{X} + \mathbf{c}$$

The Jacobian $J\left(\dfrac{y_1,y_2}{x_1,x_2}\right)$ is the determinant

$$J\left(\frac{y_1,y_2}{x_1,x_2}\right) = \begin{vmatrix} b_{11} & b_{12} \\ b_{21} & b_{22} \end{vmatrix} = |\mathbf{B}| = b_{11}b_{22} - b_{12}b_{21}$$

Assuming that $|\mathbf{B}|$ is non-zero, then X_1 and X_2 may be found in terms of Y_1 and Y_2 as in Eq. (6.3-19)

$$\mathbf{X} = \mathbf{B}^{-1}(\mathbf{Y} - \mathbf{c})$$

The joint density $f_{Y_1,Y_2}(y_1,y_2)$ may be written as

$$f_{Y_1,Y_2}(y_1,y_2) = f_{X_1,X_2}[\mathbf{B}^{-1}(\mathbf{y} - \mathbf{c})]\,\frac{1}{|\mathbf{B}|}$$

The argument $\mathbf{B}^{-1}(\mathbf{y}-\mathbf{c})$ of f_{X_1,X_2} has the two components

$$x_1 = + \frac{b_{22}}{|\mathbf{B}|}(y_1-c_1) - \frac{b_{12}}{|\mathbf{B}|}(y_2-c_2)$$

$$x_2 = - \frac{b_{21}}{|\mathbf{B}|}(y_1-c_1) + \frac{b_{11}}{|\mathbf{B}|}(y_2-c_2)$$

Thus far in this section it has been assumed that the Jacobian $J\left(\dfrac{y_1,y_2}{x_1,x_2}\right)$ exists, is non-zero, and is continuous (since the first partial derivatives $\dfrac{dy_j}{dx_i}$ are continuous). This implies that the Jacobian always has the same sign since it can only change sign by passing through zero. The situation is analogous to the one-dimensional case where the function $y = h(x)$ is either monotonically increasing or monotonically decreasing. In the one-dimensional case the Jacobian is just

$$J(y/x) = \frac{dy}{dx} \tag{7.3-10}$$

It should be clear that Eqs. (7.3-8) and (7.3-9) hold so long as the sign of the Jacobian does not change over the region of the $x_1 x_2$-plane where f_{X_1, X_2} is non-zero. What happens in any parts of the $x_1 x_2$-plane where f_{X_1, X_2} is zero is irrelevant.

Example 7.13

Let us return to the one-dimensional square-law transformation $y = x^2$ of Example 7.7 and Fig. 7.5. Note that the Jacobian is

$$J(y/x) = \frac{dy}{dx} = 2x$$

which is positive for $x > 0$, zero for $x = 0$, and negative for $x < 0$. For $x > 0$ the function is monotonically increasing, and for $x < 0$ it is monotonically decreasing. If f_X is zero for $x < 0$, then the results of Example 7.8 reduce to

$$f_Y(y) = \begin{cases} 0 & , \ y < 0 \\ \dfrac{f_X(\sqrt{y})}{2\sqrt{y}} & , \ y \ge 0 \end{cases}$$

On the other hand, if f_X is zero for $x > 0$, then

$$f_Y(y) = \begin{cases} 0 & , \ y < 0 \\ \dfrac{f_X(-\sqrt{y})}{2\sqrt{y}} & , \ y \ge 0 \end{cases}$$

Example 7.14

Consider the two-dimensional transformation

$$y_1 = x_1^2 + x_2$$
$$y_2 = x_1 + x_2^2$$

The Jacobian of this transformation is the determinant

$$J\left(\frac{y_1, y_2}{x_1, x_2}\right) = \begin{vmatrix} 2x_1 & 1 \\ 1 & 2x_2 \end{vmatrix} = 4x_1 x_2 - 1$$

The Jacobian is zero when $x_1 x_2 = 1/4$, positive when $x_1 x_2 > 1/4$, and negative when $x_1 x_2 < 1/4$ as shown in Fig. 7.10.

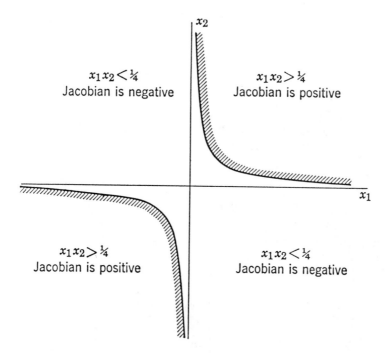

Fig. 7.10 - Regions of the $x_1 x_2$ - plane for the transformations
$$y_1 = x_1^2 + x_2 \text{ and } y_2 = x_1 + x_2^2$$

Thus the $x_1 x_2$-plane consists of three distinct regions. Although each point in the $x_1 x_2$-plane maps to a single point in the $y_1 y_2$-plane, the converse is not true; that is, a point in the $y_1 y_2$-plane may map to several distinct points in the $x_1 x_2$- plane.

In the general case where the transformations of Eq. (7.3-1) and (7.3-2) are not one-to-one, the problem may become very complicated. At least conceptually, the best procedure is to work with the distribution function as was done in Eq. (7.2-17) for the one-dimensional case. We write the joint distribution function of Y_1 and Y_2 as

$$F_{Y_1, Y_2}(y_1, y_2) = \int \int_R f_{X_1, X_2}(x_1, x_2) \, dx_1 dx_2 \qquad (7.3-11)$$

where the integration is performed over that region R of the $x_1 x_2$-plane where $Y_1 \leq y_1$ and $Y_2 \leq y_2$, that is

$$R = \{x_1, x_2 \mid (h_1(x_1, x_2) \leq y_1) \text{ and } h_2(x_1, x_2) \leq y_2)\} \qquad (7.3-12)$$

If this procedure is not too complicated to carry out, then the density function $f_{Y_1, Y_2}(y_1, y_2)$ may be obtained by differentiation. This was the procedure followed in Examples 7.7 and 7.8 in the one-dimensional case.

7.4 *Continuous Random Variables; A Special Case* - A special case that is often encountered is that where only a single function of two or more random variables is involved. For example, the transformation may be

$$Y_1 = h_1(X_1, X_2) \tag{7.4-1}$$

It is desired to find the density function of Y_1, given the joint density $f_{X_1, X_2}(x_1, x_2)$. The proper procedure to follow has already been indicated in Example 7.11 where precisely this problem was encountered. This procedure is to define a convenient random variable Y_2 by

$$Y_2 = X_2 \tag{7.4-2}$$

as was done in Example 7.11. Suppose that $h_1(x_1, x_2)$ is continuously differentiable, and such that for every pair (x_1, x_2) there corresponds one and only one value y_1. Then the Jacobian $J\left(\dfrac{y_1, y_2}{x_1, x_2}\right)$ becomes

$$J\left(\frac{y_1, y_2}{x_1, x_2}\right) = \begin{vmatrix} \dfrac{\partial y_1}{\partial x_1} & \dfrac{\partial y_1}{\partial x_2} \\ 0 & 1 \end{vmatrix} = \frac{\partial y_1}{\partial x_1} \tag{7.4-3}$$

The joint density $f_{Y_1, Y_2}(y_1, y_2)$ follows from Eq. (7.3-9) as

$$f_{Y_1, Y_2}(y_1, y_2) = f_{X_1, X_2}(x_1, x_2) \left| \frac{\partial x_1}{\partial y_1} \right| \tag{7.4-4}$$

where x_2 is just y_2 and x_1 is found as a function of y_1 and y_2 by solving Eqs. (7.4-1) and (7.4-2). Let us call this function

$$x_1 = g(y_1, y_2) \tag{7.4-5}$$

Now the density $f_{Y_1}(y_1)$ is given by

$$f_{Y_1}(y_1) = \int_S f_{X_1, X_2}[g(y_1, y_2), y_2] \left| \frac{\partial g(y_1, y_2)}{\partial y_1} \right| dy_2 \tag{7.4-6}$$

The problem of finding S, the limits of integration in y_2, may be somewhat difficult. It consists of determining where $f_{Y_1 Y_2}$ is non-zero.

In the special case where Y_1 is a linear function of X_1 and X_2, then

$$Y_1 = aX_1 + bX_2 + c \tag{7.4-7}$$

and

$$Y_2 = X_2 \tag{7.4-8}$$

where a, b, and c are constants. In this case

$$J\left(\frac{y_1, y_2}{x_1, x_2}\right) = \frac{\partial y_1}{\partial x_1} = a \qquad (7.4\text{-}9)$$

The inverse function $g(y_1, y_2)$ is just

$$x_1 = \frac{y_1 - by_2 - c}{a} \qquad (7.4\text{-}10)$$

The density $f_{Y_1}(y_1)$ of Eq. (7.4-6) becomes

$$f_{Y_1}(y_1) = \frac{1}{a}\int_S f_{X_1,X_2}\left(\frac{y_1 - by_2 - c}{a}, y_2\right) dy_2 \qquad (7.4\text{-}11)$$

Example 7.15

Consider the joint density function given by

$$f_{X_1,X_2}(x_1, x_2) = \begin{cases} 24x_1(1-x_2) &, \ 0 \leq x_1 \leq x_2, \ 0 \leq x_2 \leq 1 \\ \\ 0 &, \ \text{elsewhere} \end{cases}$$

That f_{X_1,X_2} is a density function is easily shown. Note first that it is obviously non-negative. Also

$$f_{X_2}(x_2) = \begin{cases} \int_0^{x_2} 24x_1(1-x_2)dx_1 = 12(x_2^2 - x_2^3) &, \ 0 \leq x_2 \leq 1 \\ \\ 0 &, \ \text{elsewhere} \end{cases}$$

and

$$\int_{-\infty}^{\infty} f_{X_2}(x_2)dx_2 = \int_0^1 12(x_2^2 - x_2^3)dx_2 = 1$$

The non-zero region of f_{X_1,X_2} in the $x_1 x_2$-plane is shown in Fig. 7.11(a).

We wish to find the density function of $Y_1 = X_1 + X_2$ using Eq. (7.4-11). Recall that $Y_2 = X_2$. In the $y_1 y_2$-plane, the relationship

$$y_2 = y_1 - x_1 \quad , \quad 0 \leq x_1 \leq y_2$$

holds; hence the non-zero region of f_{Y_1,Y_2} must lie between the lines $y_2 = y_1$ and $y_2 = y_1/2$ as shown in Fig. 7.11(b). Now Eq. (7.4-11) may be applied to yield

$$f_{Y_1}(y_1) = \begin{cases} \displaystyle\int_{y_1/2}^{y_1} 24(y_1-y_2)(1-y_2)dy_2 , & 0 \le y_1 \le 1 \\[2em] \displaystyle\int_{y_1/2}^{1} 24(y_1-y_2)(1-y_2)dy_2 , & 1 \le y_1 \le 2 \\[2em] 0 , & \text{elsewhere} \end{cases}$$

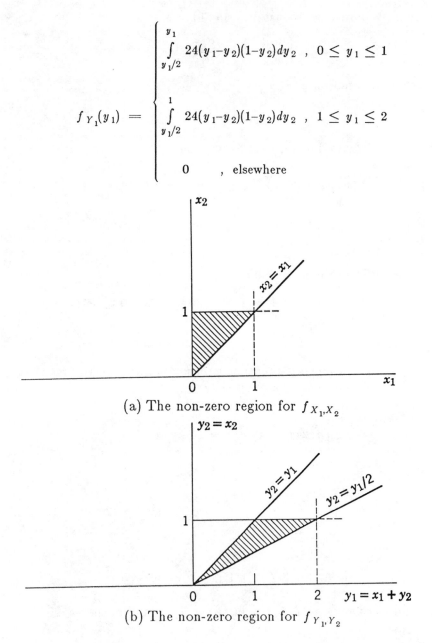

(a) The non-zero region for f_{X_1,X_2}

(b) The non-zero region for f_{Y_1,Y_2}

Fig. 7.11 - Regions of integration for Example 7.15

The integration is straightforward and the result is

$$f_{Y_1}(y_1) = \begin{cases} -2y_1^3 + 3y_1^2 , & 0 \le y_1 \le 1 \\ 2y_1^3 - 9y_1^2 + 12y_1 - 4 , & 1 \le y_1 \le 2 \\ 0 , & \text{elsewhere} \end{cases}$$

Suppose now that Eqs. (7.4-7) and (7.4-8) are satisfied and that X_1 and X_2 are *statistically independent*. This was the situation already encountered in Example 7.11. Now the joint density f_{X_1,X_2} factors into the univariate densities f_{X_1} and f_{X_2} and Eq. (7.4-11) becomes

$$f_{Y_1}(y_1) = \frac{1}{a}\int_S f_{X_1}\left(\frac{y_1-by_2-c}{a}\right) f_{X_2}(y_2)dy_2 \qquad (7.4\text{-}12)$$

Example 7.16

Let X_1 and X_2 be independent random variables which are each uniformly distributed in $[-1/2,1/2]$ as shown in Fig. 7.12(a). Find the density function of $Y_1 = X_1 + X_2$.

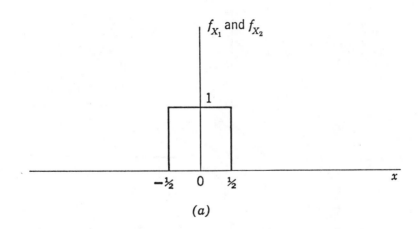

(a)

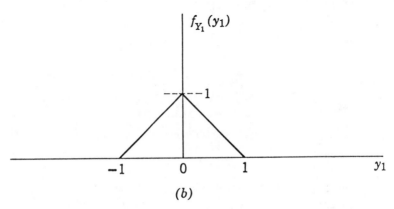

(b)

Fig. 7.12 - The density function for the sum of two independent random variables which are uniformly distributed

We will use Eq. (7.4-12), keeping in mind that $y_2 = x_2$. The only problem is to find the region in the $y_1 y_2$ -plane where $f_{Y_1 Y_2}$ is non-zero. It is immediately clear that this region must lie within the rectangle defined by $[-1 \leq y_1 \leq 1, -1/2 \leq y_2 \leq 1/2]$ since $y_2 = x_2$ has already been restricted to be non-zero in $-1/2 \leq y_2 \leq 1/2$ and since $y_2 = x_1 + x_2$ cannot be less than -1 or greater than $+1$. This rectangular region is shown in Fig. 7.13. In addition, the relationship $y_1 = x_1 + x_2 = x_1 + y_2$ further restricts the region where $f_{Y_1 Y_2}$ is non-zero to the shaded area within the rectangle of Fig. 7.13. For example, when $x_2 = y_2 = 1/2$, then $0 \leq y_1 \leq 1$, etc.

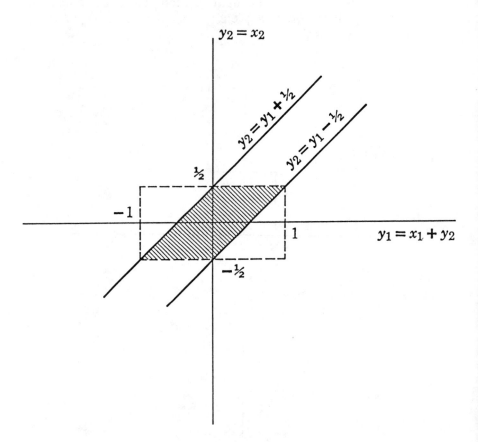

Fig. 7.13 - The non-zero region for $f_{y_1 y_2}$ of Example 7.16

Now Eq. (7.4-12) may be applied directly. The constant $a = 1$ and the integrand is unity wherever $f_{Y_1 Y_2}$ is non-zero. Two intervals of y_1 must be considered.

$-1 \leq y_1 \leq 0$.

In this interval y_2 varies from $-1/2$ at the left to the line $y_2 = y_1 + 1/2$ on the right. Therefore, in the interval $-1 \leq y_1 \leq 0$, we have

$$f_{Y_1}(y) = \int_{-1/2}^{y_1+1/2} dy_2 = 1 + y_1$$

$0 \leq y_1 \leq 1$

In this interval y_2 varies from the line $y_2 = y_1 - 1/2$ on the left to $1/2$ on the right. The density is

$$f_{Y_1}(y_1) = \int_{y_1-1/2}^{1/2} dy_2 = 1 - y_1$$

Thus, the density function $f_{Y_1}(y_1)$ is given by

$$f_{Y_1}(y_1) = \begin{cases} 1 + y_1 & , & -1 \leq y_1 \leq 0 \\ 1 - y_1 & , & 0 \leq y_1 \leq 1 \\ 0 & , & \text{elsewhere} \end{cases}$$

and is plotted in Fig. 7.12(b).

———————

Let us now consider another special case where Y_1 is the quotient of X_1 and X_2; that is,

$$Y_1 = X_1/X_2 \tag{7.4-13}$$

As before, it will be convenient to let $Y_2 = X_2$. The Jacobian of the transformation is

$$J\left(\frac{y_1, y_2}{x_1, x_2}\right) = \begin{vmatrix} \dfrac{1}{x_2} & -\dfrac{x_1}{x_2^2} \\ 0 & 1 \end{vmatrix} = \frac{1}{x_2} = \frac{1}{y_2} \tag{7.4-14}$$

It follows from Eq. (7.4-6) that the density function of Y_1 is

$$f_{Y_1}(y_1) = \int_S f_{X_1,X_2}(y_1 y_2, y_2) \, |y_2| \, dy_2 \tag{7.4-15}$$

Example 7.17

Let X_1 and X_2 be independent unit normal random variables. Equation (7.4-15) can be written immediately as

$$f_{Y_1}(y_1) = \frac{1}{2\pi} \int_{-\infty}^{\infty} e^{-y_2^2(1+y_1^2)/2} \, |y_2| \, dy_2 \, , \quad -\infty < y_1 < \infty$$

or

$$f_{Y_1}(y_1) = -\frac{1}{2\pi} \int_{-\infty}^{0} y_2 \, e^{-y_2^2(1+y_1^2)/2} \, dy_2$$

$$+ \frac{1}{2\pi} \int_{0}^{\infty} y_2 \, e^{-y_2^2(1+y_1^2)/2} \, dy_2$$

Since the integrand is a perfect differential, this expression is readily evaluated as

$$f_{Y_1}(y_1) = \frac{1}{\pi(1+y_1^2)} \quad , \quad -\infty < y_1 < \infty$$

___

Example 7.18

Consider the joint density function f_{X_1,X_2} of Example 7.15. Eq. (7.4-15) becomes

$$f_{Y_1}(y_1) = \begin{cases} 24\int_{0}^{1} y_1 y_2(1-y_2)y_2 dy_2 \ , \ 0 \leq y_1 \leq 1 \\ \\ 0 \qquad , \text{ elsewhere} \end{cases}$$

The region where f_{Y_1,Y_2} is non-zero in the $y_1 y_2$ -plane is the square with vertices $(0,0),(0,1),(1,1)$, and $(1,0)$. The integral is easily evaluated to yield

$$f_{Y_1}(y_1) = \begin{cases} 2y_1 \ , \ 0 \leq y_1 \leq 1 \\ \\ 0 \ , \text{ elsewhere} \end{cases}$$

___

Many other special cases may be treated in the same manner. As mentioned several times before, the difficulty usually lies in correctly determining that region of the $y_1 y_2$ -plane where f_{Y_1,Y_2} is non-zero.

7.5 Continuous Random Variables; The Multivariate Case - Here we treat transformations of the form

$$Y_1 = h_1(X_1, X_2, ..., X_n) = h_1(\mathbf{X})$$
$$Y_2 = h_2(X_1, X_2, ..., X_n) = h_2(\mathbf{X})$$

$$(7.5\text{-}1)$$

.

.

.

$$Y_n = h_n(X_1, X_2, ..., X_n) = h_n(\mathbf{X})$$

As in the bivariate case we assume that the transformation is one-to-one and continuously differentiable. Thus the Jacobian of the transformation is continuous. It will be assumed to be non-zero and is given by the determinant $J(\mathbf{y}/\mathbf{x})$ with ij-th term

$$a_{ij} = \frac{\partial y_i}{\partial x_j} \qquad (7.5\text{-}2)$$

where

$$J(\mathbf{y}/\mathbf{x}) = 1/J(\mathbf{x}/\mathbf{y}) \qquad (7.5\text{-}3)$$

As in the bivariate case, these conditions are only necessary and relevant whenever the joint density function $f_{\mathbf{X}}(\mathbf{x})$ of the X_i is non-zero.

Exactly as before, the joint density function of the random variables $\mathbf{Y}$ is given by

$$f_{\mathbf{Y}}(\mathbf{y}) = f_{\mathbf{X}}(\mathbf{x}) \ |J(\mathbf{x}/\mathbf{y})| \qquad (7.5\text{-}4)$$

Example 7.19

Let $f_{\mathbf{X}}(\mathbf{x})$ be the normal 3-variate density function of Problem 6.2. Let $\mathbf{y}$ be given by

$$\mathbf{y} = \mathbf{B}\,\mathbf{x}$$

where $\mathbf{B}$ is the matrix

$$\mathbf{B} = \begin{bmatrix} 1 & -1/4 & -1/2 \\ 0 & 1 & -2/7 \\ 0 & 0 & 1 \end{bmatrix}$$

In other words

$$y_1 = x_1 - \frac{1}{4} x_2 - \frac{1}{2} x_3$$

$$y_2 = x_2 - \frac{2}{7} x_3$$

$$y_3 = x_3$$

Find the density function of $\mathbf{Y}$.

The Jacobian of the transformation is given by

$$J\left(\frac{\mathbf{y}}{\mathbf{x}}\right) = |\mathbf{B}| = 1$$

We solve the x_1, x_2, and x_3 and find

$$x_1 = y_1 + \frac{1}{4} y_2 + \frac{4}{7} y_3$$

$$x_2 = y_2 + \frac{2}{7} y_3$$

$$x_3 = y_3$$

The exponent in $f_{\mathbf{X}}(\mathbf{x})$ now becomes

$$-\frac{1}{2}(2x_1^2 - x_1 x_2 + x_2^2 - 2x_1 x_3 + 4x_3^2) = -\frac{1}{2}(2y_1^2 + \frac{7}{8} y_2^2 + \frac{24}{7} y_3^2)$$

If you have solved Problem 6.2, you have found that the constant C is

$$C = \frac{\sqrt{6}}{(2\pi)^{3/2}}$$

Thus Eq. (7.5-4) may be written as

$$f_{\mathbf{Y}}(\mathbf{y}) = \frac{\sqrt{6}}{(2\pi)^{3/2}} e^{-(2y_1^2 + \frac{7}{8} y_2^2 + \frac{24}{7} y_3^2)/2}$$

or

$$f_{\mathbf{Y}}(\mathbf{y}) = \left[\frac{\sqrt{2}}{\sqrt{2\pi}} e^{-y_1^2}\right] \left[\frac{1}{\sqrt{2\pi}} \sqrt{7/8} \, e^{-7y_2^2/16}\right] \left[\frac{1}{\sqrt{2\pi}} \sqrt{24/7} \, e^{-12y_3^2/7}\right]$$

It is clear that y_1, y_2, and y_3 are independent normal random variables with zero means and variances of $\frac{1}{2}$, $\frac{8}{7}$, and $\frac{7}{24}$, respectively. The transformation $\mathbf{y} = \mathbf{B}\mathbf{x}$ is a special linear transformation which *diagonalizes* the matrix Λ^{-1} as discussed in Section 6.3. This example is a generalization to three dimensions of the problem treated in Example 6.1 in two dimensions.

This particular example could be solved directly without using Eq. (7.5-4). Since the random variables Y_1, Y_2, and Y_3 are the result of linear transformations on jointly normal random variables, they are themselves jointly normal. The covariance matrix of X_1, X_2, and X_3 is easily found from Problem 6.2 to be

$$\Lambda = \begin{bmatrix} 2/3 & 1/3 & 1/6 \\ 1/3 & 7/6 & 1/12 \\ 1/6 & 1/12 & 7/24 \end{bmatrix}$$

From the given relationship $\mathbf{Y} = \mathbf{B}\,\mathbf{X}$, it is easy to show that the Y_i are uncorrelated and, hence, independent. Their variances are easily calculated and $f_{\mathbf{Y}}(\mathbf{y})$ follows. For example,

$$E(Y_1 Y_3) = E(X_1 X_3) - \frac{1}{4} E(X_2 X_3) - \frac{1}{2} E(X_3^2)$$

$$= \frac{1}{6} - \frac{1}{4} \cdot \frac{1}{12} - \frac{1}{2} \cdot \frac{7}{24} = 0$$

and

$$E(Y_2^2) = E(X_2^2) - \frac{4}{7} E(X_2 X_3) + \frac{4}{49} E(X_3^2)$$

$$= \frac{7}{6} - \frac{4}{7} \cdot \frac{1}{12} + \frac{4}{49} \cdot \frac{7}{24} = \frac{8}{7}$$

PROBLEMS

1. If the random variable X has a density function

$$f_X(x) = \begin{cases} \dfrac{1}{\pi} , & -\dfrac{\pi}{2} < x < \dfrac{\pi}{2} \\ 0 , & \text{otherwise} \end{cases}$$

and the random variable Y is given by $Y = \cos X$, find $f_Y(y)$, the density function of Y.

2. Let the random variable X have the density function

$$f_X(x) = \begin{cases} 1/2 , & -1 < x \leq 0 \\ (2-x)/4 , & 0 < x < 2 \\ 0 , & \text{otherwise} \end{cases}$$

If the random variable Y is given by $Y = |X|$, find $f_Y(y)$, the density of Y.

3. Let X_1, X_2, and X_3 be mutually independent and uniformly distributed in the interval $[-1/2, 1/2]$. Find the density function of $Y = X_1 + X_2 + X_3$.

4. Let X_1 and X_2 be independent random variables with density functions $f_{X_1}(x_1)$ and $f_{X_2}(x_2)$, respectively where

$$f_{X_1}(x_1) = \begin{cases} 1 & , \quad 0 \le x_1 \le 1 \\ 0 & , \text{ elsewhere} \end{cases}$$

and where

$$f_{X_2}(x_2) = \begin{cases} 0 & , \quad x_2 < 0 \\ e^{-x_2} & , \quad x_2 \ge 0 \end{cases}$$

Find the density function of the sum $Y = X_1 + X_2$.

5. In Example 7.15, show that $f_{Y_1}(y_1)$ is a density function and plot it as a function of y_1.

6. Let X and Y be two independent and identically distributed random variables with a given distribution. This distribution is said to be *stable* with exponent α if $\beta(X+Y)$ has the same distribution where $\beta = 2^{-1/\alpha}$. Show that

   (a) the unit normal distribution is stable with $\alpha=2$.

   (b) the Cauchy distribution is stable with $\alpha=1$.

7. Let X be a random variable which is uniformly distributed on $(0, 2\pi)$. Find the density function of $Y = \tan X$.

8. Let $X_1, X_2, ..., X_n$ be independent random variables, each uniformly distributed on $(-1, 1)$. Find the density function of

$$Y = X_1 + X_2 + \cdots + X_n.$$

9. Let X be a normal random variable with zero mean and unit variance. Let $Y = X^2$ and show that X and Y are uncorrelated but are not independent.

10. Suppose two random variables X and Y are related by

$$Y = [\frac{X-b}{\alpha}]^c, \quad \alpha, c > 0$$

and let Y have a density function

$$f_Y(y) = e^{-y}, \quad y > 0$$

The random variable X is said to have a *Wiebull distribution*. Find the density function $f_X(x)$. Sketch for the *standard* case where $\alpha = 1$ and $b = 0$. Find the *mode* [the value of x for which $f_X(x)$ is a maximum] for $c > 1$.

11. Suppose two random variables X and Y are related by $Y = \log X$ where log is the natural logarithm and Y is $N(m, \sigma^2)$; that is, Y is normal with mean m and variance σ^2. Find the density function $f_X(x)$. Sketch for $m = 0$ and $\sigma^2 = 1$. Find the mean m_X and the variance σ_X^2. The random variable X is said to have a *log-normal* distribution. Such distributions arise in application where X can take on only non-negative values.

Chapter 8

SEQUENCES OF RANDOM VARIABLES

This chapter is concerned with the behavior of sequences $X_1, X_2, ..., X_n$ of random variables and the ways in which such sequences can converge. Some of the most important laws governing this convergence will be treated, and, in the process, it will be shown why the normal (Gaussian) distribution plays such an important role in probability theory and its applications.

A sequence of n random variables will be written in several equivalent ways:

$$\{X_n\} = \{X_n, n \geq 1\} = X_1, X_2, ..., X_n$$

It will be assumed that each X_i possesses a density function $f_i(x)$ and a distribution function $F_i(x)$ in $(-\infty, \infty)$. In general we will be interested in the ways in which $\{X_n\}$ converges as $n \to \infty$ to a random variable X with density function $f(x)$ and distribution function $F(x)$ in $(-\infty, \infty)$. The types of convergence which will be discussed are:

1) convergence in distribution (also called convergence in law),

2) convergence in probability (also called stochastic convergence),

3) almost sure convergence (also called almost certain convergence, convergence with probability one, and convergence almost everywhere),

4) convergence in the $r-th$ mean with particular emphasis on convergence in mean square.

We shall begin with a brief review of the principal notions of convergence for deterministic sequences. This material is readily available and is usually treated at an undergraduate level in the calculus sequence [Courant (Vol. 1-1937) and Hille (Vol. 1-1979)].

8.1 *Convergence in the Deterministic Case* - Let $\{s_n\} = s_1, s_2, ..., s_n$ be a sequence of numbers. The sequence is said to *converge* to the number s if

$$\lim_{n \to \infty} s_n = s \qquad (8.1\text{-}1)$$

More precisely, $\{s_n\}$ converges to s if, for each $\epsilon > 0$, there exists a positive integer $N = N(\epsilon)$ such that, for all $n \geq N$,

$$| s_n - s | < \epsilon \qquad (8.1\text{-}2)$$

Consider now a sequence of real variables $\{x_n\} = x_1, x_2, ..., x_n$ where

$$x_i = x_i(t) \text{ for } t \ \epsilon \ [a, b] \qquad (8.1\text{-}3)$$

that is, each x_i is a function of the independent variable t in the interval $a \le t \le b$. If the variable t is fixed at some particular value $t = t_0$, then the sequence

$$\{x_n(t_0)\} = x_1(t_0), \ x_2(t_0), ..., x_n(t_0) \qquad (8.1\text{-}4)$$

is just a sequence of numbers. Now if

$$\lim_{n \to \infty} x_n(t_0) = x(t_0) \qquad (8.1\text{-}5)$$

where x is a function, then the sequence $\{x_n(t_0)\}$ is said to converge to $x(t_0)$ at the *point* $t = t_0$. Such convergence is called *pointwise convergence*, and if Eq. (8.1-5) holds at every point in $[a, b]$, then $\{x_n\}$ is said to converge pointwise to x in the interval $a \le t \le b$.

A stronger condition on the sequence $\{x_n\}$ is the notion of *uniform convergence*. The sequence

$$\{x_n(t)\} = x_1(t), \ x_2(t), ..., x_n(t) \qquad (8.1\text{-}6)$$

of functions defined on $a \le t \le b$ is said to *converge uniformly* to the function $x(t)$ in $[a, b]$ if, for any $\epsilon > 0$, there exists a positive integer $N = N(\epsilon)$ such that

$$| x_n(t) - x(t) | < \epsilon \qquad (8.1\text{-}7)$$

for all $n \ge N(\epsilon)$ *and for all* $t \ \epsilon [a, b]$.

It is clear that uniform convergence in an interval $[a, b]$ implies pointwise convergence at every point in $[a, b]$. However, the converse is not true and a sequence may converge at every point in $[a, b]$ without being uniformly convergent in $[a, b]$. The idea of uniform convergence is essentially the following: Assume that the function $x(t)$ is surrounded in the interval $[a, b]$ by a circular tube of arbitrary radius $\epsilon > 0$. If a sequence $\{x_n(t)\}$ converges uniformly to $x(t)$ in $[a, b]$, then a positive integer $N = N(\epsilon)$ exists such that, for all $n \ge N$, the function $x_n(t)$ will also lie inside the tube for all $t \ \epsilon$ [a,b]; that is, all terms of the sequence from the $N\text{-}th$ term on will lie inside the ϵ -tube for all $t \epsilon [a, b]$. The tube, of course, is only two-dimensional and it will not be continuous unless $x(t)$ is continuous.

Example 8.1

Consider the following sequence

$$\{x_n(t)\} = \{1 - t^n\} \quad , \quad t \ \epsilon [-\frac{1}{2}, \frac{1}{2}]$$

For a particular $t = t_0$ in $[-1/2, 1/2]$, we have

$$\lim_{n \to \infty} [1 - t_0^n] = 1 - \lim_{n \to \infty} t_0^n = 1$$

and the sequence converges pointwise to unity in $[-1/2, 1/2]$. Is the convergence also uniform? Consider the difference

$$| x_n(t) - x(t) | = | t |^n < \epsilon$$

It is clear that $| t |^n < (\frac{2}{3})^n$ for $| t | \leq 1/2$; now choose n large enough so that $(\frac{2}{3})^n < \epsilon$ or

$$0 < \epsilon < 1 \quad , \quad n > (\ln \epsilon)/\ln 2 - \ln 3)$$
$$\epsilon \geq 1 \quad , \quad n \geq 1$$

Thus the sequence converges uniformly to unity on the interval $[a, b]$.

———

Example 8.2

Consider the sequence

$$\{x_n(t)\} = \{n^2 t e^{-nt}\}$$

It is clear that this sequence converges pointwise to zero on $[0,1]$. The function $x_n(t)$ has a maximum at $t = 1/n$ and a value at this maximum of $x_n(1/n) = n/e$ where e is the base for the natural logarithm. Form the difference

$$| x_n(t) - 0 | = n^2 t e^{-nt} < \epsilon$$

for all $t \in [0,1]$. Now choose $t = 1/n$ so that

$$| x_n(1/n) - 0 | = n/e < \epsilon$$

It is clear that for arbitrary $\epsilon > 0$ there is no $N = N(\epsilon)$ such that $N/e < \epsilon$.

The sequence does not converge uniformly in $[0,1]$. The function $x_n(t)$ is shown in Fig. 8.1 for several values of n. It is clear that an ϵ-tube cannot surround the t-axis and still contain $x_n(t)$.

———

It may be somewhat more helpful to express these notions of convergence of a function in terms of the set of points comprising its domain.

Let each of the functions x_n be defined for the points (elements) p of a set D. Then the sequence $\{x_n\} = x_1, x_2, ..., x_n$ is said to *converge pointwise* to x on a set $E \subset D$ iff. $\{x_n(p)\}$ converges for each $p \in E$; that is, if

$$\lim_{n \to \infty} x_n(p) = x(p) \quad , \quad \text{for all} \quad p \in E$$

In other words, for every $\epsilon > 0$, there is an $N = N(\epsilon,p)$ such that, for any $n \geq N$,

$$| x_n(p) - x(p) | < \epsilon \tag{8.1-8}$$

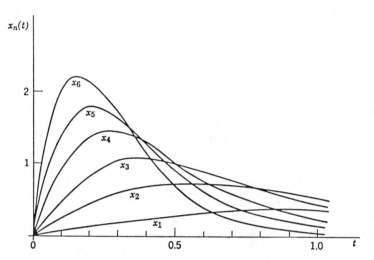

Fig. 8.1 - Nonuniform convergence

A sequence of functions $\{x_n\}$ *converges uniformly* to a function x on a set E iff, for every $\epsilon > 0$, there is an $N = N(\epsilon)$ such that, for any $n \geq N$ *and any $p \in E$,*

$$| x_n(p) - x(p) | < \epsilon \tag{8.1-9}$$

The essential difference between the two kinds of convergence is that in uniform convergence $N = N(\epsilon)$ while in pointwise convergence $N = N(\epsilon,p)$ for each $p \in E$.

Another kind of convergence that is frequently encountered, particularly in physical applications, is *convergence in the mean.* Consider again the sequence of functions $\{x_n(t)\}$ and the function $x(t)$, all defined in $a \leq t \leq b$. If

$$\lim_{n \to \infty} \int_a^b [x_n(t) - x(t)]^2 \, p(t)dt = 0$$

then $x_n(t)$ converges in the mean to $x(t)$ in the interval [a,b] and with respect to the *weighting function $p(t)$.*

For physical purposes, convergence in the mean is as useful as ordinary pointwise convergence.

We now take up the somewhat complicated subject of the convergence of random sequences.

8.2 *Convergence in Distribution* - The sequence of random variables $\{X_n\}$ is said to *converge in distribution* to the random variable X [with distribution function $F(x)$] if the distribution function $F_n(x)$ of X_n converges pointwise to $F(x)$ at all continuity points of $F(x)$; that is, if

$$\lim_{n \to \infty} F_n(x) = F(x) \qquad (8.2\text{-}1)$$

at all points of x where $F(x)$ is continuous. The relationship of Eq. (8.2-1) is sometimes written as

$$X_n \overset{d}{\to} X$$

Example 8.3

Let the random variable X_n have a distribution function $F_n(x)$ given by

$$F_n(x) = \int_{-\infty}^{x} \frac{\sqrt{n}}{\sigma\sqrt{2\pi}} \, e^{-nt^2/2\sigma^2} \, dt$$

Note that

$$\lim_{n \to \infty} F_n(x) = \begin{cases} 0 & , \; x < 0 \\ 1/2 & , \; x = 0 \\ 1 & , \; x > 0 \end{cases}$$

Thus the sequence of random variables $\{X_n\}$ converges in distribution to the random variable X with distribution function

$$F(x) = \begin{cases} 0 & , \; x < 0 \\ 1 & , \; x > 0 \end{cases}$$

It is clear that there is a discontinuity in F at $x = 0$ and that we do not have convergence at this point. However, convergence at this discontinuity is not required for X_n to converge to X in distribution.

Note that $F(x)$ is the unit step-function occurring at $x = 0$. Since $F(x)$ is a distribution function, it is right-continuous.

Example 8.4

It will now be shown that the binomial distribution $b(x;n,p)$ of Section 5.1 converges in distribution to the Poisson distribution $p(x;\lambda)$ of Section 5.2 if $np \to \lambda$. Let the random variable X_n have a distribution function

$$F_n(x) = \sum_{k=0}^{[x]} \begin{Bmatrix} n \\ k \end{Bmatrix} p_n^k (1-p_n)^{n-k} \quad , \quad x = 0,1,...,n$$

where $0 < p_n < 1$ and where $[x]$ is the integer part of x if, $x \leq n$ and $[x] = n$ if $x > n$.

We take the limit now as $n \to \infty$ in such a way that

$$\lim_{n \to \infty} np_n = \lambda$$

where λ is a fixed nonzero but finite positive number. The result is derived in detail in Section 5.2 and yields

$$\lim_{n \to \infty} F_n(x) = \sum_{k=0}^{x} \frac{\lambda^x}{x!} e^{-\lambda} \quad , \quad x = 0,1,2,...$$

which is the Poisson distribution.

The idea of convergence in distribution can be stated in another way involving the concept of the *Levy distance* between two distribution functions.

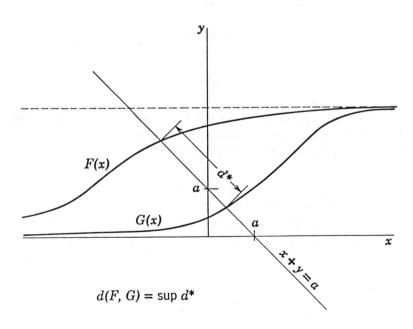

$$d(F, G) = \sup d^*$$

Fig. 8.2 - The Levy distance between two distribution functions

Let $F(x)$ and $G(x)$ be two distribution functions as shown in Fig. 8.2. Let $x+y=a$ determine a straight line. Define a distance measure $d(F,G)$ as the *least upper bound** of the distance between the intersection of $F(x)$ with $x+y=a$ and the intersection of $G(x)$ with $x+y=a$ for $a \in (-\infty,\infty)$. The measure $d(F,G)$ is called the *Levy distance* and has a number of rather obvious properties:

1) $d(F,G) = d(G,F) \geq 0$ $\hspace{3cm}$ (8.2-2)

2) $d[F(x),G(x)] = 0$ iff. $F(x) = G(x)$ $\hspace{1.5cm}$ (8.2-3)

3) $d(F,G) + d(G,H) \geq d(F,H)$ $\hspace{2cm}$ (8.2-4)

These are the usual properties of a distance measure [Feller (1966)].

A sequence of distributions $\{F_n\}$ converges to the distribution F (in the Levy sense or with the Levy metric) iff.

$$\lim_{n \to \infty} d(F_n,F) = 0 \hspace{2cm} (8.2\text{-}5)$$

It can be shown that convergence in this sense is equivalent to convergence in distribution.

Convergence in distribution is also called *convergence in law*. The *law of* X, denoted by $L(X)$, is defined as the distribution of X. If the random variables X and Y have the same distribution, it is conventional to write

$$L(X) = L(Y) \hspace{2cm} (8.2\text{-}6a)$$

or

$$X \overset{L}{=} Y \hspace{2cm} (8.2\text{-}6b)$$

In convergence in distribution it is not the values of the random variables themselves which are convergent, but, rather, the probabilities with which the random variables assume certain values.

The following result is known as *Bernstein's lemma* and is useful in proving convergence in distribution.

Let X_n, Y_n, and Z_n be random variables where

$$Z_n = X_n + Y_n \hspace{2cm} (8.2\text{-}7)$$

---

*A set R of real numbers is *bounded above* if and only if there is a real number M greater than or equal to all the elements of R. Any such number is called an *upper bound* for the set R. The number M is called the *least upper bound* (lub) if M is an upper bound and if, for every positive number ϵ, there is an element of R greater than $M-\epsilon$. The lub is also called the *supremum* (sup) of the set R.

It will be assumed that X_n and Y_n have a joint distribution $F_n(x,y)$ and X_n a distribution $F_n(x)$. Let the variance $\sigma^2(Y_n)$ go to zero as $n \to \infty$ and assume that $F_n(x)$ converges to a distribution $F(\bullet)$ with nonzero variance. Then the distribution of Z_n converges to $F(\bullet)$.

Proof: Chebychev's inequality is given in Section 4.3 and can be written as

$$P(\mid Y_n \mid \ \geq \epsilon) \leq \sigma^2(Y_n)/\epsilon^2$$

for any $\epsilon > 0$. Since $\sigma^2(Y_n) \to 0$, we have

$$\lim_{n \to \infty} P(\mid Y_n \mid \ \geq \epsilon) = 0$$

It follows from Eq. (8.2-7) that

$$P(Z_n \leq z) \leq P(X_n < z + \epsilon) + P(\mid Y_n \mid \ \geq \epsilon)$$

In the limit, we have

$$\lim_{n \to \infty} P(Z_n \leq z) \leq F(z + \epsilon) \tag{8.2-8}$$

In the same way, using Eq. (8.2-7) again,

$$P(Z_n \geq z) \geq P(X_n < z - \epsilon) + P(\mid Y_n \mid \ \geq \epsilon)$$

or

$$\lim_{n \to \infty} P(Z_n \geq z) \geq F(z - \epsilon) \tag{8.2-9}$$

It follows from Eq. (8.2-8) that, since $\epsilon > 0$ is arbitrary,

$$P(Z \leq z) \leq F(z) \tag{8.2-10}$$

and

$$P(Z \geq z) \geq F(z-0) \tag{8.2-11}$$

Thus the distribution of Z_n converges to $F(\bullet)$ or, equivalently, $\{Z_n\}$ converges in distribution to a random variable X with distribution $F(x)$. $\bullet$

8.3 *Convergence in Probability* - The sequence of random variables $\{X_n\}$ *converges in probability* to the random variable X if, for every $\epsilon > 0$,

$$\lim_{n \to \infty} P\{\mid X_n - X \mid \ \geq \epsilon\} = 0 \tag{8.3-1}$$

An equivalent statement is that, for arbitrary $\epsilon > 0$ and $\delta > 0$, there exists an integer $N = N(\epsilon, \delta)$ such that whenever $n \geq N$

$$P\{\mid X_n - X \mid \ \geq \epsilon\} < \delta \tag{8.3-2}$$

Note that this is a convergence statement about a sequence of probabilities. This type of convergence is denoted by any one of the following notations:

$$X_n \to X \quad i.p. \tag{8.3-3a}$$

$$X_n \overset{p}{\to} X \tag{8.3-3b}$$

or

$$p \lim X_n = X \tag{8.3-3c}$$

Suppose, for example, that X is a *degenerate* random variable; that is,

$$P(X = c) = 1$$

where c is a constant. Then the sequence $\{X_n\}$ converges in probability to the constant c.

Convergence in probability is sometimes called *stochastic convergence* and $\{X_n\}$ is said to converge stochastically to X.

An example of the convergence in probability of a random sequence to a constant was given in Section 5.1 and was called Bernoulli's Theorem. Other examples will be given later in this chapter when various *Weak Laws of Large Numbers* are discussed.

Example 8.5

Let the random variable X_n be distributed according to the density function given by

$$f_n(x) = \frac{n}{\pi} \frac{1}{1+n^2 x^2} \quad , \quad -\infty < x < \infty$$

and the distribution function given by

$$F_n(x) = \frac{n}{\pi} \int_{-\infty}^{x} \frac{dx}{1+n^2 x^2}$$

In the limit, the density function approaches the Dirac delta-function $\delta(x)$ of Appendix B while the distribution function approaches the unit step function $u(x)$ given by

$$\lim_{n \to \infty} F_n(x) = u(x) = \begin{cases} 0 & , \ x < 0 \\ 1 & , \ x \geq 0 \end{cases}$$

Thus we would expect that X_n converges in probability to zero, which indeed is the case as is shown by

$$\lim_{n \to \infty} P\{|X_n - 0| \geq \epsilon\} = \int_{-\infty}^{-\epsilon} \delta(x)\,dx + \int_{\epsilon}^{\infty} \delta(x)\,dx = 0$$

Example 8.6

Let the random variable X_n be given by

$$X_n = \begin{cases} 1 & \text{with probability } a^n \text{ , } a < 1 \\ 0 & \text{with probability } 1-a^n \end{cases}$$

The sequence $\{X_n\}$ converges in probability to zero since

$$P\{|X_n - 0| \geq \epsilon\} = P\{X_n = 1\} = a^n$$

and

$$\lim_{n \to \infty} a^n = 0 \quad , \quad a < 1$$

———

Let $g(x)$ be a continuous function of x in $(-\infty, \infty)$. Suppose that $\{X_n\}$ converges in probability to X. Then the sequence $\{g(X_n)\}$ converges in probability to $g(X)$.

Proof: For arbitrary $\epsilon > 0$, choose a number M such that

$$P(|X| > M) < \epsilon/2 \tag{8.3-4}$$

There exists a function $\delta(\epsilon, M)$ and three sets A, B, and C such that

$A =$ events for which $|X| \leq M$
$B =$ events for which $|X_n - X| < \delta(\epsilon, M)$
$C =$ events for which $|g(X_n) - g(X)| < \epsilon$

Then the set relation*

$$C^c \subset A^c \cup B^c$$

holds and implies that

$$P(C^c) \leq P(A^c) + P(B^c)$$

or

$$P[|g(X_n) - g(X)| \geq \epsilon] \leq P(|X| > M) + P[|X_n - X| \geq \delta(\epsilon, M)]$$

However, since $\{X_n\}$ converges in probability to X, there exists a positive integer $N(\epsilon, \delta, M)$ such that for $n > N(\epsilon, \delta, M)$,

$$P[|X_n - X| \geq \delta(\epsilon, M)] < \epsilon/2$$

Since M has been chosen to satisfy Eq. (8.3-4), it follows that $P(C^c) < \epsilon$ or

$$P[|g(X_n) - g(X)| \geq \epsilon] < \epsilon$$

———

*Recall that A^c, B^c, and C^c are the complements of sets A, B, and C.

for $n > N(\epsilon, \delta, M)$. In other words, $\{g(X_n\}$ converges in probability to $g(X)$. •

Convergence in probability implies convergence in distribution. Consider the sequence $\{X_n\} = X_1, X_2, ..., X_n$ with distribution functions $F_1(x)$, $F_2(x), ..., F_n(x)$, respectively. Assume that $\{X_n\}$ converges in probability to the random variable X with distribution function $F(x)$. Then the sequence $\{F_n(x)\}$ converges to $F(x)$ at every point of continuity of $F(x)$.

Proof: Suppose $F(x)$ is continuous at $x = x_0$. Let x_1 be a constant less than x_0; that is, $x_1 < x_0$. Define three sets A, B, and C by

$A =$ events for which $X \leq x_1$

$B =$ events for which $X_n \leq x_0$

$C =$ events for which $|X_n - X| > (x_0 - x_1)$

It is clear that $A \subset B \cup C$ and hence, that

$$P(A) \leq P(B) + P(C)$$

or

$$P(X \leq x_1) \leq P(X_n \leq x_0) + P[|X_n - X| > (x_0 - x_1)] \quad (8.3\text{-}5)$$

Since $\{X_n\}$ converges in probability to X, it follows that

$$\lim_{n \to \infty} P[|X_n - X| > (x_0 - x_1)] = 0$$

Now, in the limit, Eq. (8.3-5) becomes*

$$F(x_1) \leq \lim_{n \to \infty} P(X_n \leq x_0) \leq \lim_{n \to \infty} \inf F_n(x_0)$$

In the same way, define $x_2 > x_0$ and the new set

$D =$ events for which $X \leq x_2$

so that $D \cup C \supset B$ or

$$P(X \leq x_2) + P[|X_n - X| > (x_0 - x_1)] \geq P(X_n \leq x_0)$$

On rearranging and passing to the limit, we have

$$F(x_2) \geq \lim_{n \to \infty} P(X_n \leq x_0) - \lim_{n \to \infty} P[|X_n - X| > (x_0 - x_1)]$$

---

* A set R of real numbers is *bounded below* iff. there is a real number m less than or equal to all the elements of R. Any such number is called a *lower bound* for the set R. The number m, is called the *greatest lower bound* (glb) if m is a lower bound and if, for every positive number ϵ, there is an element of R less than $m + \epsilon$. The glb is also called the *infimum* (inf) of the set R.

or

$$F(x_2) \geq \lim_{n \to \infty} \sup F_n(x_0)$$

Thus we obtain

$$F(x_1) \leq \lim_{n \to \infty} \inf F_n(x_0) \leq \lim_{n \to \infty} \sup F_n(x_0) \leq F(x_2)$$

If $X = x_0$ is a point of continuity of $F(x)$, then

$$\lim_{x_1 \to x_0} F(x_1) = \lim_{x_2 \to x_0} F(x_2) = F(x_0)$$

and

$$\lim_{n \to \infty} F_n(x_0) = F(x_0) \bullet$$

Although convergence in probability implies convergence in distribution, the converse is not generally true. As an example, consider the case where the X_n are continuous, independent and identically distributed random variables. In this case, it is clear that, for all n,

$$d(F_n, F) = 0$$

where F is any one of the F_i. However, consider the probability $P(|X_n - X| \leq \epsilon)$ where X is an independent random variable with the same distribution as each of the X_i. It follows from Example 7.11 of Section 7.3 that this probability can be written as

$$P(|X_n - X| \leq \epsilon) = \int_{-\epsilon}^{\epsilon} \int_{-\infty}^{\infty} f(x) f(y+x) \, dx \, dy \qquad (8.3\text{-}6)$$

where $f(\bullet)$ is the density function of X and of each of the X_i. For small $\epsilon > 0$, the Mean Value Theorem allows this last equation to be given approximately by

$$P(|X_n - X| \leq \epsilon) \approx 2\epsilon \int_{-\infty}^{\infty} f^2(x) \, dx \qquad (8.3\text{-}7)$$

Thus

$$P(|X_n - X| > \epsilon) \approx 1 - 2\epsilon \int_{-\infty}^{\infty} f^2(x) \, dx \qquad (8.3\text{-}8)$$

It is clear that this last expression is independent of the index n and is not zero for arbitrary ϵ; thus $\{X_n\}$ does not converge in probability to X.

8.4 *Almost Sure Convergence* - Consider an arbitrary sample space Ω with points $\omega \, \epsilon \, \Omega$. As discussed in Section 1.7, a probability space is the triple $(\Omega, \mathbf{B}, P)$ where $\mathbf{B}$ is a Borel field on Ω and P is a probability measure defined on the events $A \in \mathbf{B}$. A random variable X or $X(\omega)$ was defined in Section 2.1 as a real and single-valued function whose domain is Ω and which is $\mathbf{B}$-measurable; that is, for every real number x,

$$\{\omega \in \Omega \mid X(\omega) \leq x\} \in \mathbf{B}$$

Thus the random variable X is more correctly written as $X(\omega)$ although the explicit functional dependence on the underlying space is usually suppressed in the interest of simplicity of notation.

A sequence $\{X_n\}$ of random variables is said to converge *almost surely* (a.s.) to the random variable X if the sequence of numbers $\{X_n(\omega)\}$ converges to $X(\omega)$ for all sample points $\omega \in \Omega$ except possibly for those belonging to a set of probability zero; that is,*

$$P\{\omega \mid \lim_{n \to \infty} X_n(\omega) = X(\omega)\} = 1 \qquad (8.4\text{-}1)$$

An equivalent statement is that, with probability 1, for arbitrary $\epsilon > 0$, there exists an integer $N = N(\epsilon, \omega)$ such that, for $n \geq N(\epsilon, \omega)$,

$$|X_n(\omega) - X(\omega)| < \epsilon \qquad (8.4\text{-}2)$$

Note that two random variables X_n and X are said to be *equal almost surely* if

$$P(X_n \neq X) = 0 \qquad (8.4\text{-}3)$$

This type of convergence is also called *almost certain* (a.c.) convergence, convergence *with probability one*, and convergence *almost everywhere* (a.e.). Various notations are used including

$$X_n \overset{a.s.}{\to} X \quad or \quad X_n \to X \; (a.s.) \qquad (8.4\text{-}4a)$$

$$X_n \overset{a.c.}{\to} X \quad or \quad X_n \to X \; (a.c.) \qquad (8.4\text{-}4b)$$

$$X_n \overset{prob\,1}{\to} X \quad or \quad X_n \to X \; (prob\,1) \qquad (8.4\text{-}4c)$$

$$X_n \overset{a.e.}{\to} X \quad or \quad X_n \to X \; (a.e.) \qquad (8.4\text{-}4d)$$

---

*Recall that Eq. (8.4-1) will usually be written as

$$P\{\lim_{n \to \infty} X_n(\omega) = X(\omega)\} = 1$$

This type of convergence is a probability statement about a convergent sequence of functions. Another condition equivalent to Eq. (8.4-1) is that, for arbitrary $\epsilon > 0$ and $\delta > 0$, there exists an integer $N = N(\epsilon, \delta)$ such that

$$P\left\{ \bigcap_{n=N}^{\infty} [\omega \mid X_n(\omega) - X(\omega)] < \epsilon] \right\} \geq 1 - \delta \qquad (8.4\text{-}5)$$

Almost sure convergence is more or less the strongest kind of convergence and implies various equalities among random variables.

Example 8.7

Let the sample space Ω be the closed unit interval $[0,1]$ and define a random variable $X_n(\omega)$ by

$$X_n(\omega) = \left\{ \begin{array}{ll} 1 & , \ \omega \leq 1/n \\ 0 & , \ \omega > 1/n \end{array} \right.$$

Let $X(\omega)$ be the degenerate random variable zero in $0 \leq \omega \leq 1$. Now for each pair of numbers (a,b) satisfying

$$0 \leq a \leq b \leq 1$$

define an associated probability

$$P(a \leq \omega \leq b) = b - a$$

Using this probability measure, we have

$$P\{X_n(\omega) = 1\} = 1/n$$

and

$$P\{X_n(\omega) = 0\} = 1 - 1/n$$

For every $\omega \neq 0$ and arbitrary $\epsilon > 0$, there exists an $N = N(\epsilon)$ such that for $n > N$

$$|X_n(\omega) - X(\omega)| < \epsilon$$

In other words,

$$\lim_{n \to \infty} X_n(\omega) = X(\omega) \quad , \quad \omega \neq 0$$

and

$$P\{ \lim_{n \to \infty} X_n(\omega) = X(\omega)\} = P\{\Omega - \{0\}\} = 1$$

where $\{0\}$ is the set consisting of the element zero. It follows from Eq. (8.4-1) that $\{X_n(\omega)\}$ converges almost surely to $X(\omega)$.

Convergence almost surely implies convergence in probability.

Proof: Refer to Eq. (8.4-5) which is a necessary and sufficient condition for convergence almost surely. Now consider any sequence of events $\{A_n\} = A_0, A_1, ..., A_n, ...$ It is clear that

$$\bigcap_{n=0}^{\infty} A_n \subset A_0 \qquad (8.4-6)$$

Thus it must follow that

$$P\{\bigcap_{n=0}^{\infty} A_n\} \leq P\{A_0\} \qquad (8.4-7)$$

Equation (8.4-5) is equivalent to

$$P\{\bigcap_{n=N}^{\infty} [\omega \mid \mid X_n(\omega) - X(\omega) \mid \geq \epsilon]\} < \delta \qquad (8.4-8)$$

But Eq. (8.4-7) implies that

$$P\{\omega \mid \mid X_n(\omega) - X(\omega) \mid \geq \epsilon\}$$

$$(8.4-9)$$

$$= P\{\mid X_n(\omega) - X(\omega) \mid \geq \epsilon\} < \delta$$

which was to be proved. •

Although convergence almost surely implies convergence in probability, the converse is not true as the following example shows.

Example 8.8

Let the sample space Ω be the closed unit interval $[0,1]$. Also for each pair of numbers (a, b) satisfying

$$0 \leq a \leq b \leq 1$$

define an associated probability by

$$P(a \leq \omega \leq b) = b - a$$

Let us now define a new random variable by

$$Y_{ij}(\omega) = \begin{cases} 1, & (j-1)/i < \omega \leq j/i \\ 0, & \text{elsewhere} \end{cases}$$

for $i, j = 1, 2, ...$. Also define n by

$$n = \frac{i(i-1)}{2} + j \ ,$$

set $X_n(\omega) = Y_{ij}(\omega)$, and let $X(\omega) = 0$. It is clear that, for $\epsilon \geq 1$,

$$P\{\mid X_n(\omega) - X(\omega) \mid > \epsilon\} = 0$$

and, for $0 < \epsilon < 1$,

$$P \{ \mid X_n (\omega) - X(\omega) \mid > \epsilon \} = P \{X_n (\omega) = 1\} = 1/i$$

We have the inequality

$$n < \frac{i(i-1)}{2} + \frac{3}{2} i + \frac{1}{2} = \frac{(i+1)^2}{2}$$

Thus, $i > \sqrt{2n} - 1$ and

$$\lim_{n \to \infty} P \{X_n (\omega) = 1\} \leq \lim_{n \to \infty} \frac{1}{\sqrt{2n} - 1} = 0$$

In other words $\{X_n (\omega)\}$ converges in probability to $X(\omega)$. However, $\{X_n (\omega)\}$ does not converge almost surely to $X(\omega)$. Consider any point $\omega \in \Omega$ such that $\omega \neq 0$. For any $N > 0$, there is an $n > N$ such that $X_n (\omega) = 1$. Thus $X_n (\omega)$ does not converge to $X(\omega)$ for any $\omega \in \Omega$ and

$$P \{ \lim_{n \to \infty} X_n (\omega) = X(\omega)\} = 0$$

8.5 *Convergence in r–th Mean* - The sequence of random variables $\{X_n\}$ is said to *converge in r–th mean* $(r > 0)$ to the random variable X if $E (\mid X_n \mid^r) < \infty$ and $E (\mid X \mid^r) < \infty$ and if

$$\lim_{n \to \infty} E (\mid X_n - X \mid^r) = 0 \qquad (8.5\text{-}1)$$

This last expression is sometimes written as

$$X_n \overset{r}{\to} X \qquad (8.5\text{-}2a)$$

and sometimes as

$$X_n \to X \quad (mean \; r) \qquad (8.5\text{-}2b)$$

When $r = 2$, $\{X_n\}$ is said to *converge in mean-square* or *in the mean of order two.* In some texts the notation

$$l.i.m._{n \to \infty} X_n = X \qquad (8.5\text{-}3a)$$

or

$$X_n \to X \quad (m.s.) \qquad (8.5\text{-}3b)$$

is used for mean-square convergence.

Convergence in the r–th mean implies convergence in probability.

Proof: It follows from the Bienayme-Chebychev theorem of Eq. (4.3-1) that

$$P \{ \mid X_n - X \mid \geq \epsilon \} \leq \frac{E \{X_n - X \mid^r \}}{\epsilon^r} \qquad (8.5\text{-}4)$$

for arbitrary $\epsilon > 0$. If $\{X_n\}$ converges in the $r-th$ mean to X, then it is clear from Eq. (8.5-1) that $\{X_n\}$ converges to X in probability since

$$\lim_{n \to \infty} P\{\mid X_n - X \mid \geq \epsilon\} = 0 \qquad \bullet \qquad (8.5\text{-}5)$$

Convergence in the mean of order r also implies convergence in the mean of all lower orders; that is, suppose that

$$X_n \overset{r}{\to} X \qquad (8.5\text{-}6)$$

and s is such that

$$0 < s < r \qquad (8.5\text{-}7)$$

then

$$X_n \overset{s}{\to} X \qquad (8.5\text{-}8)$$

Proof: It is shown in Appendix E that

$$h(p) = \{E(\mid X \mid^p)\}^{1/p} \qquad (8.5\text{-}9)$$

is a non-decreasing function of p for $p > 0$. On taking logarithms, we obtain, for $0 < s < r$

$$\frac{\log E\{\mid X_n - X \mid^s\}}{s} \leq \frac{\log E\{\mid X_n - X \mid^r\}}{r} \qquad (8.5\text{-}10)$$

If $\{X_n\}$ converges in $r-th$ mean to X, then the right side of this last equation approaches minus infinity in the limit. Thus the left side also approaches minus infinity in the limit, which implies that $X_n \overset{s}{\to} X$. $\bullet$

Although convergence in $r-th$ mean implies convergence in probability, the converse is not generally true as the following example shows.

Example 8.9

Refer to Example 8.5 of Section 8.3 and consider the sequence of random variables described there. Form

$$E\{\mid X_n - 0 \mid^2\} = E\{\mid X_n \mid^2\}$$

$$= \frac{n}{\pi} \int_{-\infty}^{\infty} \frac{x^2}{1 + n^2 x^2} \, dx$$

This integral does not exist even for infinite n and the sequence does not converge in $r-th$ mean even though it converges in probability.

Convergence in the r-th mean does not generally imply convergence almost surely. Consider the following example.

Example 8.10

Let $\{X_n\}$ be a sequence of independent random variables defined by

$$X_n = \begin{cases} n^{1/2r} & , \text{ with probability } 1/n \\ 0 & , \text{ with probability } 1-1/n \end{cases}$$

and let X be the degenerate random variable which is always zero. We have, for $r > 0$,

$$E\{\,|\,X_n\,|^r\,\} = (n^{1/2r})^r \; n^{-1} = n^{-1/2}$$

Thus

$$\lim_{n \to \infty} E\{\,|\,X_n\,|^r\,\} = 0$$

and $\{X_n\}$ converges in r-th mean to $X \equiv 0$.

Consider values of n such that $N \leq n \leq M$. We can write

$$P\{\text{all } X_n = 0 \text{ for } N \leq n \leq M\} = \prod_{n=N}^{M} (1 - 1/n)$$

or, for $\epsilon > 0$ *and for arbitrary* N,*

$$P\{\bigcap_{n=N}^{\infty} [\omega\,|\,X_n(\omega) < \epsilon]\} = \lim_{M \to \infty} \prod_{n=N}^{M} (1-1/n) = 0$$

Thus $\{X_n\}$ does not converge almost surely to $X \equiv 0$.

———

On the other hand, neither does convergence almost surely implies convergence in the r-th mean as the following example shows.

Example 8.11

Let $\{X_n\}$ be a sequence of independent random variables defined by

———

*We have [Jolley (1961), p. 198, no. 1067]

$$\prod_{n=2}^{\infty} (1-1/n) = 0$$

and, hence, for arbitrary N,

$$\prod_{n=N}^{\infty} (1-1/n) = 0$$

$$X_n = \begin{cases} e^n & , \text{ with probability } 1/n^2 \\ 0 & , \text{ with probability } 1-1/n^2 \end{cases}$$

and let X be the degenerate random variable which is always zero. For any $r > 0$, we have

$$\lim_{n \to \infty} E\{|X_n|^r\} = \lim_{n \to \infty} e^{nr}/n^2 = \infty$$

and $\{X_n\}$ does not converge to zero in the r-th mean for any r. However, we can write

$$P\{\text{all } X_n = 0 \text{ for } N \leq n \leq M\} = \prod_{n=N}^{M} (1-1/n^2)$$

or, for $\epsilon > 0$,

$$P\{\bigcap_{n=N}^{\infty} [\omega \mid X_n(\omega) < \epsilon]\} = \lim_{M \to \infty} \prod_{n=N}^{M} (1-1/n^2)$$

We have [Jolley (1961), p. 198, No. 1068], for $N \geq 2$

$$(1/2) \leq \prod_{n=N}^{\infty} (1 - 1/n^2) \leq 1$$

It follows that the product

$$\lim_{M \to \infty} \prod_{n=N}^{M} (1-1/n^2)$$

converges to a number between one-half and unity and that this number approaches unity as $N \to \infty$, that is,

$$\lim_{N \to \infty} \prod_{n=N}^{\infty} (1-1/n^2) = 1$$

Thus, as shown by Eq. (8.4-5), the sequence $\{X_n\}$ converges almost surely to X.

8.6 *Relations Among Types of Convergence* - It was shown in Section 8.3 that convergence in probability implies convergence in distribution but that the converse is not true. It was shown in Section 8.4 that convergence almost surely implies convergence in probability but that the converse is not true as shown by Example 8.8. In Section 8.5, it was shown that convergence in r-th mean implies convergence in the mean to all orders lower than r. In addition such convergence implies convergence in probability but, as shown in Example 8.9, the converse is not true. It was also shown by Examples 8.10 and 8.11 that convergence in r-th mean does not imply convergence almost surely and convergence almost surely does not imply convergence in the r-th mean. All of the foregoing relationships and implications are shown diagrammatically in Fig. 8.3.

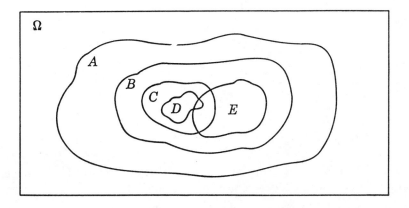

Ω - Set of all sequences $\{X_n\}$ of random variables
A - Set of all sequences convergent in distribution
B - Set of all sequences convergent in probability
C - Set of all sequences convergent in s th mean ($s > 0$)
D - Set of all sequences convergent in r th mean ($r > s$)
E - Set of all sequences convergent almost surely

$$A \supset B \supset C \supset D \quad \text{and} \quad B \supset E$$

or

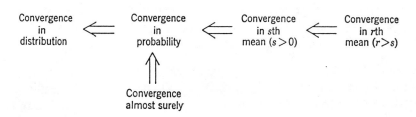

Fig. 8.3 - Relationships among types of convergence of sequences
of random variables

It is clear that convergence in distribution is the weakest, followed by convergence in probability. Although convergence in distribution does not, in general, imply convergence in probability, the implication does exist when $\{X_n\}$ converges in probability to a constant.

Proof: Let the sequence $\{X_n\}$ converge in distribution to a constant c : that is, let

$$\lim_{n \to \infty} F_n(x) = F(x)$$

where

$$F(x) = \begin{cases} 0 & , \ x < c \\ 1 & , \ x \geq c \end{cases}$$

Then, for any $\epsilon > 0$,

$$P\{\ |\ X_n - c\ |\ \leq \epsilon\} = F_n(c+\epsilon) - F_n(c-\epsilon-0)$$

and

$$\lim_{n\to\infty} P\{\ |\ X_n - c\ |\ \leq \epsilon\} = F(c+\epsilon) - F(c-\epsilon-0) = 1$$

Thus $\{X_n\}$ converges in probability to the constant c; the converse has been proved already in Section 8.3. •

In order for convergence in probability to imply convergence in r-th mean, it is necessary to impose some boundedness condition. Such a condition is provided by the *Lebesque dominated - convergence theorem:* Let $\{Y_n\}$ be a sequence of random variables which converges almost surely to the random variable Y, and let a random variable Z exist such that $|\ Y_n\ | \leq Z$ for all n. If $E(Z) < \infty$, then

$$\lim_{n\to\infty} E(Y_n) = E(Y)$$

and

$$\lim_{n\to\infty} E[\ |\ Y_n - Y\ |] = 0$$

For a proof of this theorem, refer to [Munroe (1953), p. 186]. •

It is now possible to go from convergence almost surely to convergence in the r-th mean by adding suitable restrictions.

Proof: Let $Y_n = |\ X_n - X\ |^r$, let $Y \equiv 0$, and assume that $X_n \overset{a.s.}{\to} X$. If $E(Z) < \infty$ where $Z \geq |\ X_n - X\ |^r$, then, by the Lebesgue dominated - convergence theorem,

$$\lim_{n\to\infty} E[\ |\ X_n - X\ |^r] = 0$$

or $\{X_n\}$ converges in r-th mean to X. •

8.7 *Limit Theorems on Sums of Random Variables* - In Section 5.1, a simple limit theorem known as *Bernoulli's Theorem* was developed. This theorem stated that the relative frequency with which an event occurs in a sequence of n Bernoulli trials converges in probability to the probability p of occurrence in a single trial. It was stated then that this theorem was a special case of a limit theorem known as the *Law of Large Numbers*. We now proceed to derive a sequence of general limit theorems including the Law of Large Numbers.

The limit theorems that will be treated in the next sections are:

1. The Weak Law of Large Numbers
   a. Bernoulli's Theorem

2. The Strong Law of Large Numbers

3. The Central Limit Theorem

Before taking up these theorems, let us consider briefly some results involving sequences of events.

Let $\{X_n\}$ be a sequence of random variables possessing a joint distribution. Let $\{E_n\} = E_1, E_2, ..., E_n$ be a *sequence of events* with associated probabilities $P(E_n) = p_n$ and $P(E_n^c) = 1-p_n$. Define the *indicator function* of Section 2.1 by

$$I_{E_i}(\bullet) = \begin{cases} 1 & \text{if } E_i \text{ occurs} \\ 0 & \text{if } E_i^c \text{ occurs} \end{cases} \tag{8.7-1}$$

and let

$$X_i = I_{E_i}(\bullet) \tag{8.7-2}$$

Then it is true that

$$P\{X_n = 1\} = P\{E_n\} = p_n = 1 - P\{X_n = 0\} \tag{8.7-3}$$

The *Borel - Cantelli lemmas* can now be stated and proved.

Lemma 1: if the events E_i are independent, the probability of occurrence of an infinite number of them is 0 or 1 depending on whether Σp_n converges or not.

Lemma 2: Regardless of whether or not the events are independent, the convergence of Σp_n implies, with probability 1, that only a finite number occur.

Proof: Assume that Σp_n converges and let q be the probability that an infinite number of events occur. Let $q_{M,N}$ be the probability that *at least one* of the events $E_M, E_{M+1}, ..., E_N$ occurs, so that

$$q = \lim_{M \to \infty} \left[\lim_{N \to \infty} q_{M,N} \right] \tag{8.7-4}$$

It follows from Eq. (1.7-9) that

$$q_{M,N} \le p_M + p_{M+1} + \cdots + p_N \tag{8.7-5}$$

Hence, Eq. (8.7-4) becomes

$$q \le \lim_{M \to \infty} \sum_{n=M}^{\infty} p_n = 0 \tag{8.7-6}$$

since Σp_n converges. This proves Lemma 2 since the E_i have not yet been assumed to be independent. It also proves the first part of Lemma 1.

Assume now that the E_i are *independent* and that Σp_n diverges. Let $r_{M,N}$ be the probability that *none* of the events $E_M, E_{M+1}, ..., E_N$ occur. This probability is given by

$$r_{M,N} = 1 - q_{M,N} = \prod_{n=M}^{N} (1-p_n) \qquad (8.7\text{-}7)$$

On taking the limit, we have

$$\lim_{N \to \infty} r_{M,N} = 1 - \lim_{n \to \infty} q_{M,N} = \lim_{N \to \infty} \prod_{n=M}^{N} (1-p_n) = 0 \quad (8.7\text{-}8)$$

or finally,

$$\lim_{n \to \infty} q_{M,N} = q = 1 \qquad (8.7\text{-}9)$$

and the second part of Lemma 1 is proved. •

Consider now a sequence $\{X_n\}$ of random variables, a sequence $\{c_n\}$ of constants, and the event E_n that $|X_n| > c_n$. Let p_n be the probability

$$p_n = P(|X_n| > c_n) \qquad (8.7\text{-}10)$$

It follows from the Borel-Cantelli lemmas that the probability of the occurrence of an infinite number of the events E_n is zero if Σp_n converges. If the X_n are independent and Σp_n diverges, the probability of the occurrence of an infinite number of the events is unity.

Consider sequences of Bernoulli trials (see Section 5.1) where each trial admits the outcomes S (success) and F (failure) with probabilities $P(S)=p$ and $P(F)=1-p$. Let Ω be the space consisting of all infinite sequences of S s and F s. Let $\omega \in \Omega$ and let ω_i be the $i-th$ letter in ω, that is,

$$\omega = (\omega_1, \omega_2, ..., \omega_n, ...)$$

Let S_i be the event of a success on the $i-th$ trial and define the random variable X_i by

$$X_i = X_i(\omega) = I_{S_i}(\omega) = \begin{cases} 1 & \text{if } \omega_i = S \\ 0 & \text{if } \omega_i = F \end{cases} \qquad (8.7\text{-}11)$$

Now the number of successes in n tosses is the random variable

$$Y_n = \sum_{u=1}^{n} X_i \qquad (8.7\text{-}12)$$

and the relative frequency of success is

$$Y_n/n = \sum_{i=1}^{n} X_i/n \qquad (8.7\text{-}13)$$

It was shown in Section 5.1 that Bernoulli's Theorem requires that this relative frequency converge in probability to p; that is, $Y_n/n \overset{p}{\to} p$.

Let us now define an event E by

$$E = \{\omega \mid \frac{X_1 + X_2 + ... + X_n}{n} \xrightarrow{p} p \} \tag{8.7-14}$$

Note that whether or not $\omega \in E$ *does not depend* on the first k of the X_i. In other words, we also have, for any $k \geq 1$,

$$E = \{\omega \mid \frac{X_k + X_{k+1} + ... + X_n}{n} \xrightarrow{p} p \} \tag{8.7-15}$$

Any set having this property - that whether or not $\omega \in E$ does not depend on the first k coordinates of ω, no matter how large is k - is called a *tail event*. More precisely: "Let $(\Omega, \mathbf{B}, \mathbf{P})$ be a probability space. Let $\mathbf{X} = (X_1, X_2, ..., X_n, ...)$ be any random sequence. A set (event) $E \in \mathbf{B}(\mathbf{X})$ will be called a *tail event* if $E \in \mathbf{B}(X_k, X_{k+1}, ...)$ for all k. Equivalently, let $\mathbf{F}$ be the Borel field given by

$$\mathbf{F} = \bigcap_{k=1}^{\infty} \mathbf{B}(X_k, X_{k+1}, ...) \tag{8.7-16}$$

Then $\mathbf{F}$ is called the *tail Borel field* (or *tail σ-field*) and any set $E \in \mathbf{F}$ is a *tail event*." An unusual theorem involving tail events is the *Kolmogorov Zero-One Law:* Let $X_1, X_2, ..., X_n, ...$ be a sequence of independent random variables. If $E \in \mathbf{F}(\mathbf{X})$, a tail σ-field, then $P(E)$ is either zero or one.

Proof: see [Neveu (1965) p. 128]. •

Example 8.12

Consider any specific sequence s of k Bernoulli trials and let the event F_n be the occurrence of this sequence beginning with the n-th trial; that is

$$F_n = \{\omega \mid (\omega_n, \omega_{n+1}, ..., \omega_{n+k-1}) = s \}$$

where $0 < P(\omega_i) < 1$. Then the probability that the sequence occurs infinitely often (for an infinite number of n) in an infinite sequence of Bernoulli trials is equal to unity.

Proof: Define the events A_i by

$$A_1 = \{\omega \mid (\omega_1, \omega_2, ..., \omega_k) = s \}$$
$$A_2 = \{\omega \mid (\omega_{k+1}, \omega_{k+2}, ..., \omega_{2k}) = s \}$$
$$A_3 = \cdots$$

Now the events A_i are independent and

$$\left\{ \begin{array}{l} event\ that\ F_n\ occurs \\ infinitely\ often \end{array} \right\} \supset \left\{ \begin{array}{l} event\ that\ A_n\ occurs \\ for\ an\ infinite\ number\ of\ n \end{array} \right\}$$

Also we have

$$P(A_n) = P(A_1) > 0$$

and

$$\sum_{n=1}^{\infty} P(A_n) = \infty$$

It now follows from the Borel-Cantelli lemmas that

$$P \left\{ \begin{array}{l} event\ that\ A_n\ occurs \\ for\ an\ infinite\ number\ of\ n \end{array} \right\} = 1$$

which implies

$$P \left\{ \begin{array}{l} event\ that\ F_n\ occurs \\ infinitely\ often \end{array} \right\} = 1 \quad \bullet$$

For the most part, two types of limit theorems will be encountered in probability theory:

Weak limit theorems: Given a sequence $\{X_n\}$ of random variables, it is shown that there is a random variable X for which

$$\lim_{n \to \infty} P\{ |X_n - X| \geq \epsilon \} = 0$$

that is, $X_n \overset{p}{\to} X$. Examples which will be considered here are the Weak Law of Large Numbers and the Central Limit Theorem.

Strong limit theorems: Given a sequence $\{X_n\}$ of random variables, it is shown that there is a random variable X for which

$$P\{ \lim_{n \to \infty} X_n = X \} = 1$$

that is, $X_n \overset{a.s.}{\to} X$. An example which will be considered is the Strong Law of Large Numbers.

We consider the Weak Law of Large Numbers in the following section.

8.8 The Weak Law of Large Numbers - Let $\{X_n\} = X_1, X_2, ..., X_n$ be a sequence of random variables with means $E(X_i) = m_i$, $i = 1, 2, ..., n$, and variances $E[(X_i - m_i)^2] = \sigma_i^2$, $i = 1, 2, ..., n$. Define a new sequence of random variables $\{Y_n\}$ by the $n-th\ partial\ sum$

$$Y_n = \sum_{i=1}^{n} X_i \quad , \quad n = 1,2,\dots \qquad (8.8\text{-}1)$$

The random variable Y_n has mean μ_n given by

$$\mu_n = E(Y_n) = \sum_{i=1}^{n} m_i \qquad (8.8\text{-}2)$$

and variance s_n^2

$$s_n^2 = E[(Y_n - \mu_n)^2] \qquad (8.8\text{-}3)$$

Define the n-th $arithmetic$ $average$ A_n by

$$A_n = Y_n / n \qquad (8.8\text{-}4)$$

which has a mean

$$E(A_n) = \mu_n / n \qquad (8.8\text{-}5)$$

and a variance

$$E[(A_n - \mu_n/n)^2] = s_n^2 / n^2 \qquad (8.8\text{-}6)$$

A number of convergence statements can be made about the sequences $\{X_n\}$ and $\{A_n\}$, and these constitute the *Weak Law of Large Numbers*.

Suppose the variance of X_n approaches zero in the limit; that is,

$$\lim_{n \to \infty} \sigma_n^2 = 0 \qquad (8.8\text{-}7)$$

The Chebychev inequality of Section 4.3 can be applied to X_n to write, for $\epsilon > 0$,

$$P(|X_n - m_n| \geq \epsilon) \leq \sigma_n^2 / \epsilon^2 \qquad (8.8\text{-}8)$$

In the limit, this last expression states that the sequence $\{X_n - m_n\}$ converges in probability to zero:

$$\lim_{n \to \infty} P(|X_n - m_n| \geq \epsilon) = 0 \qquad (8.8\text{-}9)$$

In the same way, if

$$\lim_{n \to \infty} s_n^2 / n^2 = 0 \qquad (8.8\text{-}10)$$

then

$$\lim_{n \to \infty} P(|A_n - \mu_n/n| \geq \epsilon) = 0 \qquad (8.8\text{-}11)$$

and the sequence $\{A_n - \mu_n/n\}$ converges in probability to zero.

If the X_n are *mutually independent* random variables, then Eq. (8.8-10) is satisfied if the variances of the X_n are bounded; that is, if there exists a positive number M such that $\sigma_n^2 \leq M$ for all n. In this case, Eq. (8.8-10) can be written as

$$\lim_{n \to \infty} s_n^2/n^2 = \lim_{n \to \infty} \sum_{i=1}^{n} \sigma_i^2/n^2 \leq \lim_{n \to \infty} M/n = 0 \qquad (8.8\text{-}12)$$

This case obviously includes the special situation where the X_n are mutually independent identically distributed random variables with the same mean m and variance σ^2. Now Eq. (8.8-10) becomes

$$\lim_{n \to \infty} s_n^2/n^2 = \lim_{n \to \infty} \sigma^2/n = 0 \qquad (8.8\text{-}13)$$

This last result involving independent and identically distributed random variables is often called the Law of Large Numbers.

At this point, it is clear that Bernoulli's Theorem discussed in Section 5.1 is a special case of the Weak Law of Large Numbers. Let E_i be the event that a success occurs on the i-th Bernoulli trial. Then the set $\{X_n\}$ is a sequence of independent identically distributed random variables given by

$$X_i = I_{E_i}(\bullet) \qquad (8.8\text{-}14)$$

where $I_{E_i}(\bullet)$ is the indicator of Section 2.1. Each X_i has a mean of $m_i = p$ and a variance of $\sigma_i^2 = p(1-p)$. Hence the n-th partial sum Y_n is the number of successes in n trials with mean $\mu_n = np$ and variance $s_n^2 = np(1-p)$. The n-th arithmetic average A_n is the relative number of successes in n trials with mean $\mu_n/n = p$ and variance $s_n^2/n^2 = p(1-p)/n$. Thus Eq. (8.8-7) is satisfied and A_n converges in probability to p as stated in Section 5.1.

8.9 *The Strong Law of Large Numbers* - The Weak Law of Large Numbers developed in the last section considered conditions under which the mean of a sequence of random variables converges *in probability* to the arithmetic average of the mean of each random variable. As mentioned in Section 8.7, the Strong Law of Large Numbers has to do with the convergence of such sequences *almost surely*.

More specifically, consider the sequence of random variables $\{X_n\}$ and define the n-th partial sum Y_n by Eq. (8.8-1). Let $\{a_n\}$ and $\{b_n\}$ be two sequences of real numbers such that

$$\lim_{n \to \infty} b_n = \infty \qquad (8.9\text{-}1)$$

The sequence $\{Y_n\}$ is said to be *stable in probability* if

$$\left[\frac{Y_n}{b_n} - a_n \right] \xrightarrow{p} 0 \qquad (8.9\text{-}2)$$

and to be *stable almost surely* if

$$\left[\frac{Y_n}{b_n} - a_n \right] \xrightarrow{a.s.} 0 \qquad (8.9\text{-}3)$$

The sequence $\{X_n\}$ is said to obey the *Weak Law of Large Numbers* with respect to the constants $\{b_n\}$ if a sequence $\{a_n\}$ exists such that Eq. (8.9-2) holds. The sequence $\{X_n\}$ obeys the *Strong Law of Large Numbers* with respect to the constants $\{b_n\}$ if a sequence $\{a_n\}$ exists such that Eq. (8.9-3) holds. The last section was concerned with the identification of the $\{a_n\}$ and $\{b_n\}$ for Eq. (8.9-2); this identification was $\{a_n\} = \{\frac{\mu_n}{n}\}$ and $\{b_n\} = \{n\}$.

We shall not prove any of the various forms of the Strong Law of Large Numbers but shall list some of the more important ones. Proofs will be found in Doob (1953), Feller (1966), and Breiman (1968). It is clear that any sequence which obeys a Strong Law of Large Numbers also obeys the corresponding Weak Law of Large Numbers since convergence almost surely implies convergence in probability.

Kolmogorov's First Strong Law of Large Numbers

Let the $\{X_n\}$ be a sequence of *independent* random variables with finite means $E\{X_i\} = m_i$ and variances $Var\ X_i = \sigma_i^2$. Let the sequence $\{a_n\}$ be given by

$$a_n = \frac{E\{Y_n\}}{b_n} = \frac{\mu_n}{b_n} \qquad (8.9\text{-}4)$$

and assume that the series

$$\sum_{i=1}^{\infty} \sigma_i^2/b_i^2 < \infty \qquad (8.9\text{-}5)$$

where $\{b_n\}$ is any sequence of positive constants obeying Eq. (8.9-1). Then $\{X_n\}$ obeys the Strong Law of Large Numbers with respect to the b_n.

Borel's Strong Law of Large Numbers

Let $\{X_n\}$ be defined by Eq. (8.8-14) so that we consider a sequence of Bernoulli trials with probability of success p and probability of failure $q = 1-p$. Let $a_n = p$ and $b_n = n$. Then the sequence $\{X_n\}$ obeys the Strong Law of Large Numbers; that is, the relative frequency of success converges almost surely to p or

$$\left\{ \frac{Y_n}{n} - p \right\} \xrightarrow{a.s.} 0$$

Kolmogorov's Second Strong Law of Large Numbers

Let $\{X_n\}$ be a sequence of *independent and identically distributed* random variables with the common distribution function $F(x)$. Let $b_n = n$ and let $a_n = a$ where a is the mean of X_i; that is

$$a = \int_{-\infty}^{\infty} x \ dF(x)$$

A *necessary* and *sufficient* condition for $\{X_n\}$ to satisfy the Strong Law of Large Numbers is that a, the common mean, exists [i.e. $|E(X_i)| < \infty$].

We proceed now to consider briefly some of the remarkable properties of the Central Limit Theorem. This theorem states that, under reasonably general conditions, sums of random variables tend to be distributed asymptotically normally.

8.10 *The Central Limit Theorem* - We continue with the notation of the previous two sections and restrict our attention to cases where the sequence $\{X_n\}$ consists of independent random variables with means $E\{X_i\} = m_i$ and variances $Var\ X_i = \sigma_i^2$. We work with the normalized random variable Z_n where

$$Z_n = \frac{Y_n - \mu_n}{s_n} \qquad (8.10\text{-}1)$$

and Y_n, μ_n, and s_n are given by Eqs. (8.8-1), (8.8-2), and (8.8-3), respectively. The new random variable Z_n has a mean of zero and a variance of unity. We now consider several forms of the Central Limit Theorem under which Z_n converges in distribution to the unit normal distribution.

Lindeberg-Levy Theorem -- *Identically Distributed Random Variables*

The simplest case to consider is that where the X_i are not only *independent* but also *identically distributed* with common mean m and variance σ^2. Now Eq. (8.10-1) reduces to

$$Z_n = \frac{Y_n - nm}{\sqrt{n}\ \sigma} = \sum_{i=1}^{n} \frac{X_i - m}{\sqrt{n}\ \sigma} \qquad (8.10\text{-}2)$$

We now show that, in the limit, Z_n has a unit normal distribution.

Proof: Define a moment generating function $M(t)$ by

$$M(t) = E\{e^{t(X_i - m)/\sigma}\} \qquad (8.10\text{-}3)$$

Let $M_n(t)$ be the moment generating function for Z_n so that

$$M_n(t) = E\{e^{tZ_n}\} = E\left\{\exp\left[\frac{t}{\sqrt{n}}\sum_{i=1}^{n}\frac{X_i-m}{\sigma}\right]\right\} \quad (8.10\text{-}4)$$

Since the X_i are independent, this last expression can be written as

$$M_n(t) = \prod_{i=1}^{n} M\left(\frac{t}{\sqrt{n}}\right) = \left[M\left(\frac{t}{\sqrt{n}}\right)\right]^n \quad (8.10\text{-}5)$$

The power series expansion of $M\left(\frac{t}{\sqrt{n}}\right)$ yields

$$M\left(\frac{t}{\sqrt{n}}\right) = \quad (8.10\text{-}6)$$

$$1 + \frac{t}{\sqrt{n}}u_1 + \frac{t^2}{2n}u_2 + \cdots + \frac{t^n}{n!(n)^{n/2}}u_n + \cdots$$

where

$$u_n = E\left\{\left[\frac{X_n-m}{\sigma}\right]^n\right\} \quad (8.10\text{-}7)$$

and

$$u_1 = 0 \quad , \quad u_2 = 1$$

Now Eq. (8.10-6) can be approximated *for large n* as*

$$M\left(\frac{t}{\sqrt{n}}\right) = 1 + \frac{t^2}{2n} + \left[o\left(\frac{t^2}{n}\right)\right] \approx e^{t^2/2n} \quad (8.10\text{-}8)$$

and Eq. (8.10-5) becomes, in the limit

$$\lim_{n\to\infty} M_n(t) = \left[e^{t^2/2n}\right]^n = e^{t^2/2} \quad (8.10\text{-}9)$$

which is the moment generating function of a unit normal random variable. Thus Z_n converges in distribution to a unit normal random variable. •

We now wish to generalize these results to the case where the X_i are still independent but not necessarily identically distributed. It will be necessary to add some condition to insure that each X_i has a negligible influence on the asymptotic distribution.

Liapounov Theorem -- Bounded Variance

Let the X_i be independent with means m_i and variances σ_i^2 and assume that, for $\delta > 0$,

---

* The function $o\left(\frac{t^2}{n}\right)$ has the property that

$$\lim_{n\to\infty} \frac{n}{t^2} o\left(\frac{t^2}{n}\right) = 0$$

$$\lim_{n \to \infty} s_n^{-2-\delta} \sum_{i=1}^{n} E\left\{(X_i - m_i)^{2+\delta}\right\} = 0 \qquad (8.10\text{-}10)$$

Then the random variable

$$Z_n = \frac{Y_n - \mu_n}{s_n} \qquad (8.10\text{-}11)$$

has a distribution which is asymptotically unit normal.

Proof: see Cramer (1946) and Breiman (1968). ●

The Liapounov Theorem is sometimes stated in a slightly different form [Cramer (1946) p. 215], and it is shown that a *sufficient* set of conditions for Z_n to converge in distribution to a unit normal random variable is that two numbers l and L exist such that

$$\sigma_i^2 > l > 0 \qquad , \quad \text{all } i \qquad (8.10\text{-}12)$$

and

$$E\left\{|X_i - m_i|^3\right\} < L \qquad , \quad \text{all } i \qquad (8.10\text{-}13)$$

These are sufficient conditions to insure that no one term of the series dominates.

It should be kept in mind that the preceding results concerning the Central Limit Theorem have involved convergence in distribution. This does not necessarily mean that the density function of Z_n converges to a unit normal density function. If the X_i are continuous random variables, then, under some regularity conditions, the density function of Z_n does converge to the unit normal density. If the X_i are discrete random variables, the situation is more complicated and the envelope of the density function of Z_n may converge to the unit normal density.

PROBLEMS

1. Let $\{X_n\}$ be a sequence of independent identically distributed random variables with common distribution function $F(x)$ given by

$$F(x) = \begin{cases} 0 & , \ x \leq 0 \\ x & , \ 0 < x \leq 1 \\ 1 & , \ x > 1 \end{cases}$$

Define two sequences $\{Y_n\}$ and $\{Z_n\}$ by

$$Y_n = max(X_1, X_2, ..., X_n)$$

and

$$Z_n = n(1 - Y_n)$$

Show that $\{Z_n\}$ converges in distribution to a random variable Z with distribution function

$$F(z) = 1 - e^{-z}$$

2. Define a random variable X_i by

$$Z_i = k + N_i \quad , \quad i=1,2,...,n$$

where k is a constant and the N_i are independent, identically distributed random variables with zero mean. Show that the sequence $\{Y_n\}$ where

$$Y_n = \frac{1}{n} \sum_{i=1}^{n} X_i \quad , \quad n=1,2,...$$

converges in probability to k.

3. Let $\{X_n\}$ be a sequence of random variables which converge in probability to the random variable X. Let the density function $f_n(x)$ of X_n be such that, when $n > N$,

$$f_n(x) = 0 \quad , \quad \text{for } |x| > x_0$$

Show that $\{X_n\}$ converges mean-square to X.

APPENDIX A

THE RIEMANN-STIELTJES INTEGRAL

A.1 The Riemann Integral- A fundamental problem in the elementary calculus is the determination of the area under a positive curve $f(x)$ in some interval $[a,b]$* such as is shown in Fig. A.1. The usual way to approach this problem is to subdivide the interval into m segments $[x_i - x_{i-1}]$ and to form the sums

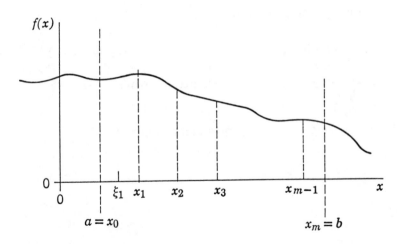

Fig. A.1 - Determination of the Area Under a Curve

$$\sum_{i=1}^{m} f(\xi_i)[x_i - x_{i-1}] \qquad (A.1\text{-}1)$$

where ξ_i is a point in the ith interval. In other words, any arbitrary set of ξ_i and x_i produce a partition of $[a,b]$ given by

$$a = x_o < \xi_1 < x_1 < \xi_2 < x_2 < ... < \xi_m < x_m = b \qquad (A.1\text{-}2)$$

---

* The notation $[a,b]$ will be used to denote a *closed interval;* that is, one which includes the end points a and b. The *open interval* will be denoted by (a,b).

The sum represented by Eq. (A.1-1) has a different value for each partition of $[a,b]$, of course, but each such value is an approximation to the area between $f(x)$ and the x-axis in the interval $[a,b]$. It is intuitively apparent that, if $f(x)$ is reasonably well behaved, then the sum becomes a better and better approximation to the area as the number m increases and as the length $[x_i - x_{i-1}]$ of the longest interval decreases. For sufficiently well behaved $f(x)$ a unique limit exists which is the *Riemann definite integral* of elementary calculus. More precisely, the Riemann integral of $f(x)$ with respect to x over the interval $[a,b]$ is defined as

$$\int_a^b f(x)\,dx = \lim_{\max|x_i - x_{i-1}| \to 0} \sum_{i=1}^m f(\xi_i)[x_i - x_{i-1}] \qquad (A.1\text{-}3)$$

for any arbitrary sequence of partitions of the form of Eq. (A.1-2).

At this point a word about notation is in order. It is clear that, in ordinary summations,

$$\sum_{i=1}^n a_i = \sum_{j=1}^n a_j = a_1 + a_2 + \dots + a_n \qquad (A.1\text{-}4)$$

The symbols i and j are each called an "index of summation" and the sum does not depend on the notation used for this index. In the same way

$$\int_a^b f(x)\,dx = \int_a^b f(y)\,dy \qquad (A.1\text{-}5)$$

and the definite integral depends of a, b and f but not on the "variables of integration" x and y.

A.2 - Functions of Bounded Variation. A function $g(x)$ is called *non-decreasing* in $[a,b]$ if $g(x_2) \geq g(x_1)$ for any pair of values x_1 and x_2 in $[a,b]$ for which $x_2 > x_1$. It is called *increasing* if $g(x_2) > g(x_1)$. Similarly, $g(x)$ is called *non-increasing* in $[a,b]$ if $g(x_2) \leq g(x_1)$. It is called *decreasing* if $g(x_2) < g(x_1)$. Such functions are also called *monotonic*. More precisely, they are called *monotone non-decreasing* or *monotone increasing*, etc.

A function $g(x)$ is of *bounded variation* [Apostal (1974)] in $[a,b]$ if and only if there exists a number M such that

$$\sum_{i=1}^m |g(x_i) - g(x_{i-1})| < M \qquad (A.2\text{-}1)$$

for all partitions

$$a = x_o < x_1 < x_2 < \ldots < x_m = b \qquad \text{(A.2-2)}$$

of the interval. Alternately, $g(x)$ is of bounded variation if and only if it can be written in the form

$$g(x) = g_1(x) - g_2(x) \qquad \text{(A.2-3)}$$

where the functions $g_1(x)$ and $g_2(x)$ are bounded and non-decreasing in $[a,b]$. If the function $g(x)$ is bounded and has a finite number of relative maxima and minima, and discontinuities in the finite open interval (a,b), then it is of bounded variation in (a,b). This last statement is known as the *Dirichlet Conditions* [Apostal (1974)] and turns out to be important in the treatment of Fourier series.

Where realizable signals are concerned, if $g(t)$ is of bounded variation in some interval, then it is bounded and its high frequency components must make a limited total contribution to its frequency spectrum.

A.3 - The Riemann-Stieltjes Integral. The Riemann-Stieltjes integral [Apostal (1974)] of the function $f(x)$ with respect to the function $g(x)$ over the interval $[a,b]$ is defined as

$$\int_a^b f(x)\,dg(x) = \lim_{\max|x_i - x_{i-1}| \to 0} \sum_{i=1}^m f(\xi_i)[g(x_i) - g(x_{i-1})] \qquad \text{(A.3-1)}$$

for any arbitrary sequence of partitions of the form of Eq. (A.1-2). A sufficient condition for the existence of the limit given by Eq. (A.3-1) is that $g(x)$ be of *bounded variation* and $f(x)$ be *continuous* on $[a,b]$ (or that $f(x)$ be of bounded variation and $g(x)$ be continuous). It is apparent that, for the case where

$$g(x) = \alpha x, \qquad \text{(A.3-2)}$$

the Riemann-Stieltjes integral reduces to the ordinary Riemann integral.

Some idea of the geometric meaning of the Riemann-Stieltjes integral may be obtained by referring to Fig. A.2. It is apparent that the Riemann-Stieltjes sum

$$\sum_{i=1}^m f(\xi_i)[g(x_i) - g(x_{i-1})] \qquad \text{(A.3-3)}$$

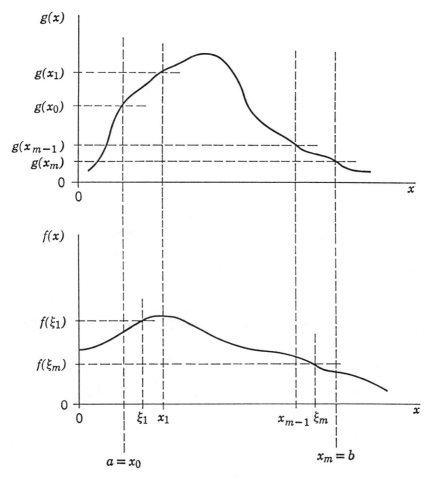

Fig. A.2 - The Reimann-Stieltjes Integral

is not an approximation to the area under $f(x)$ except when $g(x)$ is a straight line through the origin with unit slope. Rather the sum is weighted by the values that $g(x)$ assumes in the interval $[a,b]$. As mentioned in the preceding paragraph, one of the functions f or g need not be particularly well behaved as far as continuity is concerned but only of bounded variation. For example, suppose that $u_-(x)$ is the unit step function given by

$$u_-(x) = \begin{cases} 0 & x < 0 \\ 1 & x \geq 0 \end{cases} \qquad \text{(A.3-4)}$$

Then the Riemann-Stieltjes integral may be used to write the sum

$$\sum_k \alpha_k f(k) = \int_{-\infty}^{\infty} f(x) d \sum_k \alpha_k u_-(x-k) \qquad \text{(A.3-5)}$$

In particular, let $f(x)$ be a density function which may be either discrete or continuous or mixed. Its cumulative distribution function $F(a)$ is

$$F(a) = P\{x \leq a\} = \int_{-\infty}^{a} dF(x) \qquad \text{(A.3-6)}$$

and the expression

$$\int_{-\infty}^{\infty} dF(x) = 1 \qquad \text{(A.3-7)}$$

replaces both Eq. (2.9-12) and Eq. (2.9-24).

The Riemann-Stieltjes integral, if it exists, has a number of properties analogous to those of the Riemann integral as well as some for which there is no obvious analogy:

1)

$$\int_{a}^{b} (\alpha_1 f_1 + \alpha_2 f_2) dg = \alpha_1 \int_{a}^{b} f_1 dg + \alpha_2 \int_{a}^{b} f_2 dg \qquad \text{(A.3-8)}$$

2)

$$\int_{a}^{b} f d(\alpha_1 g_1 + \alpha_2 g_2) = \alpha_1 \int_{a}^{b} f \, dg_1 + \alpha_2 \int_{a}^{b} f \, dg_2 \qquad \text{(A.3-9)}$$

3)

$$\int_{a}^{b} f dg = -\int_{b}^{a} f \, dg \qquad \text{(A.3-10)}$$

4)

$$\int_{a}^{b} f \, dg = 0 \qquad \text{(A.3-11)}$$

if g is constant in $[a,b]$.

5)

$$\int_{a}^{b} f \, dg = f [g(b) - g(a)] \qquad \text{(A.3-12)}$$

if f is constant in $[a,b]$.

6)

$$\int_{a}^{b} f \, dg = f(a)[k - g(a)] + f(b)[g(b) - k] \qquad \text{(A.3-13)}$$

if g is the constant k in (a,b) and f is continuous at a and b.

7)

$$\int_a^b f(x)dg(x) = f(b)g(b) - f(a)g(a) - \int_a^b g(x)df(x) \quad \text{(A.3-14)}$$

from integration by parts.

8)

$$\int_a^b f(x)dg(x) = \int_a^b f(x)g(x)dx \quad \text{(A.3-15)}$$

if $g(x)$ is continuously differentiable on $[a,b]$.

APPENDIX B

THE DIRAC DELTA FUNCTION

B.1 Step Functions - As mentioned in Appendix A, it is often convenient to introduce the concept of a *step function* [Apostol (1974)], a function which changes its value only on a discrete set of discontinuities. The commonest step function considered is the *unit-step function* which may be defined in at least three ways:

1) symmetrical unit-step function u (x)

$$u(x) = \begin{cases} 0 & x < 0 \\ 1/2 & x = 0 \\ 1 & x > 0 \end{cases} \qquad (B.1\text{-}1)$$

2) asymmetrical unit-step function u_(x)

$$u_-(x) = \begin{cases} 0 & x < 0 \\ 1 & x \geq 0 \end{cases} \qquad (B.1\text{-}2)$$

3) asymmetrical unit-step function u_+(x)

$$u_+(x) = \begin{cases} 0 & x \leq 0 \\ 1 & x > 0 \end{cases} \qquad (B.1\text{-}3)$$

Each of these is illustrated in Fig. B.1. It is apparent that $u_+(x)$ is continuous on the left and the $u_-(x)$ is continuous on the right at $x = 0$.

B.2 The Dirac Delta Function - The Dirac delta function $\delta(x)$ [Dirac (1958)] is of considerable utility in physical applications even though it is mathematically pathological. It may be defined in terms of its "sampling" property with regard to an arbitrary function $f(x)$, for $a < b$:

$$\int_a^b f(x)\delta(x-k)dx = \begin{cases} 0 \text{ for } k < a \text{ or } k > b \\ 1/2 \ f(k+0) \text{ for } k = a \\ 1/2 \ f(k-0) \text{ for } k = b \\ 1/2[f(k-0)+f(k+0)] \text{ for } a < k < b \end{cases} \qquad (B.2\text{-}1)$$

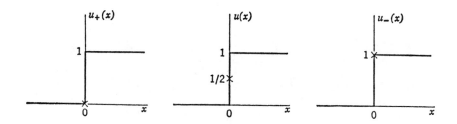

Fig. B.1 - Unit-step functions

In the usual case where (a,b) is $(-\infty,\infty)$ and where $f(x)$ is continuous, Eq. (B.2-1) becomes simply

$$\int_{-\infty}^{\infty} f(x)\delta(x-k)\,dx = f(k) \tag{B.2-2}$$

Note that $\delta(x)$ is not a function by the usual definitions since Eq. (B.2-2) implies that

$$\delta(x) = 0 , \ x \neq 0 \tag{B.2-3}$$

and that

$$\int_{-\infty}^{\infty} \delta(x)\,dx = \int_{-\epsilon}^{\epsilon} \delta(x)\,dx = 1 \tag{B.2-4}$$

for any $\epsilon > 0$, which are not consistent relationships. A more general approach to the definition of "function" can be made through the *theory of distributions* [Lighthill (1958)] in order to justify definitions such as Eq. (B.2-1). For our purposes we will regard the delta function as a formal representation of the operation defined by Eq. (B.2-1) and will not attempt to further justify its use.

A comparison of Eqs. (A.3-5) and (B.2-2) indicates that the use of the delta function could be avoided by employing the Riemann-Stieltjes integral:

$$\int_{-\infty}^{\infty} f(x)\,du\,(x-k) = \int_{-\infty}^{\infty} f(x)\delta(x-k)\,dx = f(k) \qquad \text{(B.2-5)}$$

This relationship also suggest that, in a formal sense, the delta function can be regarded as the "derivative" of the unit-step function:

$$\frac{d}{dx}\,u\,(x) = \delta(x) \qquad \text{(B.2-6)}$$

In the same way, the r-th "derivative" $\delta^{(r)}(x)$ of the delta function is defined for $a < b$ by

$$\int_{a}^{b} f(x)\delta^{(r)}(x-k)\,dx = \begin{cases} 0 & \text{for } k < a \text{ and } k > b \\ (1/2)(-1)^r\, f^{(r)}(k+0) & \text{for } k = a \\ (1/2)(-1)^r\, f^{(r)}(k-0) & \text{for } k = b \\ (1/2)(-1)^r\,[f^{(r)}(k-0)+f^{(r)}(k+0)] & \text{for } a < k < b \end{cases} \qquad \text{(B.2-7)}$$

where $f^{(r)}(x)$ is the r-th derivative of $f(x)$ with respect to x. Again, for the infinite interval and continuous $f(x)$, this becomes

$$\int_{-\infty}^{\infty} f(x)\delta^{(r)}(x-k)\,dx = (-1)^r\, f^{(r)}(k) \qquad \text{(B.2-8)}$$

A number of formal relationships involving the delta function may be obtained directly from the defining relationship of Eq. (B.2-1). For example:

1) $\delta(x) = \delta(-x)$ \qquad\qquad\qquad\qquad\qquad (B.2-9)

2) $\delta(\alpha x) = \dfrac{1}{\alpha}\,\delta(x)$, $\alpha > 0$ \qquad\qquad\qquad (B.2-10)

3) $f(x)\delta(x-k) = \dfrac{1}{2}[f(k-0) + f(k+0)]\delta(x-k)$ \qquad (B.2-11)

4) $x\,\delta(x) = 0$ \qquad\qquad\qquad\qquad\qquad\qquad (B.2-12)

5) $\delta(x^2-\alpha^2) = \dfrac{1}{2\alpha}\,[\delta(x-\alpha)+\delta(x+\alpha)]$, $\alpha > 0$ \qquad (B.2-13)

6) $\int_{-\infty}^{\infty} \delta(\alpha-x)\delta(x-\beta)\,dx = \delta(\alpha-\beta)$ \qquad\qquad (B.2-14)

APPENDIX C

INTERCHANGE OF ORDER IN DIFFERENTIATION AND INTEGRATION

Let $f(x)$ be a function integrable on the interval $[a,b]$; that is,

$$\int_a^x f(y)\, dy = F(x) < \infty \quad , \quad a \le x \le b \tag{C-1}$$

If $f(x)$ is continuous at some point x_0 in $[a,b]$, then $F(x)$ has a derivative at $x = x_0$ and this derivative is

$$\frac{dF(x)}{dx} \Big|_{x=x_0} = f(x_0) \tag{C-2}$$

Let $f(x,y)$ and $\dfrac{\partial f(x,y)}{\partial y}$ be continuous in a rectangle given by $a \le x \le b$ and $\alpha \le y \le \beta$. Suppose the function $F(y)$ exists where

$$F(y) = \int_a^b f(x,y)\, dx \tag{C-3}$$

Then $F(y)$ has a derivative with respect to y in the interval $[\alpha, \beta]$ and it is given by

$$\frac{dF(y)}{dy} = \int_a^b \frac{\partial}{\partial y}[f(x,y)]\, dx \tag{C-4}$$

Let $g(x,y)$ be a function such that the integral

$$G(y) = \int_{-\infty}^\infty g(x,y)\, dx \tag{C-5}$$

exists. Then the derivative of $G(y)$ at $y = y_0$ is given by

$$\frac{dg(y)}{dy} \Big|_{y=y_0} = \int_{-\infty}^\infty \left[\frac{\partial g(x,y)}{\partial y} \right]_{y=y_0} dx \tag{C-6}$$

if there exists a function $h(x)$ such that, for some non-zero ϵ_0,

$$\frac{\partial g(x,y)}{\partial y} \Big|_{y=y_0+\epsilon} \le h(x) \tag{C-7}$$

for all ϵ such that $|\epsilon| \le \epsilon_0$, and if

$$\int_{-\infty}^{\infty} h(x)\ dx\ <\infty \tag{C-8}$$

Let $[a,b]$ and $[c,d]$ be arbitrary intervals and define the definite integral

$$I = \int_{a}^{b}\int_{c}^{d} h(x,y)\ dy\ dx \tag{C-9}$$

Any one of the following three conditions is sufficient for the interchange of the order of integration in Eq. (C-9):

$$A = \int_{a}^{b}\int_{c}^{d} |h(x,y)|\ dy\ dx\ <\infty \tag{C-10}$$

$$B = \int_{a}^{b} dx \int_{c}^{d} |h(x,y)|\ dy\ <\infty \tag{C-11}$$

$$C = \int_{c}^{d} dy \int_{a}^{b} |h(x,y)|\ dx\ <\infty \tag{C-12}$$

If these last three equations are satisfied then $A=B=C$. If any one of the three is satisfied, then

$$\int_{a}^{b}\int_{c}^{d} h(x,y)\ dy\ dx = \int_{c}^{d}\int_{a}^{b} h(x,y)\ dx\ dy \tag{C-13}$$

APPENDIX D

SOME ELEMENTS OF MATRIX THEORY

D.1 Basic Concepts - A *matrix* is a collection of mn elements arranged in m rows and n columns. In talking about matrices, we shall consistently consider first rows and then columns so that a matrix of *order* $m \times n$ will always have m rows and n columns. Each element of the matrix will be denoted by a lower case letter with two subscripts, the first to identify the row and the second the column. Thus a_{ij} is the element in the i-th row and the j-th column. A matrix will be represented by a bold-faced capital letter such as $\mathbf{A}$ or $\mathbf{B}$ or by the symbol $[a_{ij}]$; that is,

$$\mathbf{A} = [a_{ij}] \tag{D.1-1}$$

A *square matrix* is one with the same number of rows as columns. Such a matrix is called *symmetric* if

$$a_{ij} = a_{ji} \tag{D.1-2}$$

The *zero matrix* or *null matrix* $\mathbf{O}$ has all of its elements equal to zero. The *identity matrix* $\mathbf{I}$ is a square matrix such that

$$a_{ij} = \begin{cases} 1 & , \ i=j \\ 0 & , \ i \neq j \end{cases} \tag{D.1-3}$$

The set of elements where $i=j$ is called the *main diagonal* of the matrix. An identity matrix has 1's for the elements a_{ii} of its main diagonal and 0's for the other elements.

Addition and subtraction - If $\mathbf{A} = [a_{ij}]$ and $\mathbf{B} = [b_{ij}]$ are two matrices of the *same* order $m \times n$, then the sum is an $m \times n$ matrix $\mathbf{C} = [c_{ij}]$ with elements

$$c_{ij} = a_{ij} + b_{ij} \tag{D.1-4}$$

It is clear from this last equation that

$$\mathbf{A} + \mathbf{O} = \mathbf{A} \tag{D.1-5}$$

$$\mathbf{A} + \mathbf{B} = \mathbf{B} + \mathbf{A} \tag{D.1-6}$$

and

$$\mathbf{A} + (\mathbf{B}+\mathbf{C}) = (\mathbf{A}+\mathbf{B}) + \mathbf{C} = \mathbf{A} + \mathbf{B} + \mathbf{C} \tag{D.1-7}$$

In the same way the difference $\mathbf{A}$–$\mathbf{B}$ between two $m \times n$ matrices is the matrix $\mathbf{C}$ with elements

$$c_{ij} = a_{ij} - b_{ij} \qquad \text{(D.1-8)}$$

Scalar multiplication - Let α be any scalar. Then the product $\alpha\mathbf{A} = \alpha[a_{ij}]$ is a matrix of the same order as $\mathbf{A}$ with elements αa_{ij}. It follows that

$$\alpha(\mathbf{A}+\mathbf{B}) = \alpha\mathbf{A} + \alpha\mathbf{B} \qquad \text{(D.1-9)}$$

and

$$(\alpha+\beta)\mathbf{A} = \alpha\mathbf{A} + \beta\mathbf{A} \qquad \text{(D.1-10)}$$

Matrix multiplication - Two matrices $\mathbf{A}$ and $\mathbf{B}$ may be multiplied to form the product $\mathbf{A}\,\mathbf{B}$ iff. the number of *columns* of $\mathbf{A}$ is equal to the number of *rows* of $\mathbf{B}$. Let $\mathbf{A}$ be an $m \times n$ matrix with elements a_{ij} and let $\mathbf{B}$ be an $n \times p$ matrix with elements b_{ij}. Then $\mathbf{A}\,\mathbf{B} = \mathbf{C}$ is an $m \times p$ matrix with elements c_{ij} given by

$$c_{ij} = \sum_{k=1}^{n} a_{ik}\, b_{kj} \qquad \text{(D.1-11)}$$

Note that, in general, the commutative law does *not* hold for matrix multiplication and

$$\mathbf{A}\,\mathbf{B} \neq \mathbf{B}\,\mathbf{A} \quad \text{(in general)} \qquad \text{(D.1-12)}$$

However, if the matrices are compatible, the associative and distributive laws hold so that

$$(\mathbf{AB})\mathbf{C} = \mathbf{A}(\mathbf{BC}) = \mathbf{ABC} \qquad \text{(D.1-13)}$$

$$\mathbf{A}(\mathbf{B}+\mathbf{C}) = \mathbf{A}\,\mathbf{B} + \mathbf{A}\,\mathbf{C} \qquad \text{(D.1-14)}$$

and

$$(\mathbf{A}+\mathbf{B})(\mathbf{C}+\mathbf{D}) = \mathbf{A}(\mathbf{C}+\mathbf{D}) + \mathbf{B}(\mathbf{C}+\mathbf{D}) \qquad \text{(D.1-15)}$$

If $\mathbf{I}$ is an $m \times m$ identity matrix and $\mathbf{A}$ is an arbitrary $m \times m$ matrix, then

$$\mathbf{I}\,\mathbf{A} = \mathbf{A}\,\mathbf{I} = \mathbf{A} \qquad \text{(D.1-16)}$$

Transpose of a matrix - The transpose of a matrix $\mathbf{A}$ with elements a_{jk} is denoted by $\mathbf{A}^T$ and has elements $a_{jk}^T = a_{kj}$; that is, the rows and columns are interchanged. It is easy to show that

$$(\mathbf{A}\,\mathbf{B})^T = \mathbf{B}^T\,\mathbf{A}^T \qquad \text{(D.1-17)}$$

and

$$(\mathbf{A}\,\mathbf{B}\,\mathbf{C})^T = \mathbf{C}^T\,\mathbf{B}^T\,\mathbf{A}^T \qquad \text{(D.1-18)}$$

Vectors - A *one* $\times n$ matrix has a single row and is sometimes called a *row vector* ; an $n \times one$ matrix has a single column and is sometimes called a *column vector*. Note that the transpose of a row vector is a column vector and vice-versa. Let $\mathbf{R}$ be a *one* $\times n$ row vector and let $\mathbf{C}$ be an $m \times one$ column vector. The product $\mathbf{R} \, \mathbf{C}$ exists only if $m = n$ and is a *one* $\times one$ matrix or *scalar*. Note that $\mathbf{C} \, \mathbf{R}$ is an $m \times n$ matrix.

Inverse of a matrix - The inverse $\mathbf{A}^{-1}$ of a matrix $\mathbf{A}$ is a matrix such that

$$\mathbf{A} \, \mathbf{A}^{-1} = \mathbf{A}^{-1} \, \mathbf{A} = \mathbf{I} \qquad \text{(D.1-19)}$$

where $\mathbf{I}$ is the identity matrix. Note that $\mathbf{A}$ must be *square* to have an inverse, and this inverse is unique. A square matrix whose inverse does not exist is said to be *singular*. The matrix $\mathbf{A}$ is singular iff. there exists a row vector $\mathbf{R}$ which is non-zero and such that

$$\mathbf{R} \, \mathbf{A} = 0 \qquad \text{(D.1-20)}$$

If the matrix $\mathbf{A}$ is singular, then its transpose $\mathbf{A}^T$ is also singular.

Determinant - The determinant $|\mathbf{A}|$ of an $m \times m$ square nonsingular matrix $\mathbf{A} = [a_{ij}]$ is a real-valued function of the elements a_{ij} and is defined as

$$|\mathbf{A}| = |a_{ij}| = \sum (\pm \, a_{1i} \, a_{2j} \dots a_{mp}) \qquad \text{(D.1-21)}$$

where the summation is taken over all permutations $(i, j, \dots, p)$ of $(1, 2, \dots, m)$ with a plus sign if $(i, j, \dots, p)$ is an even permutation and a minus sign if it is an odd permutation. The number of permutations is $m!$.

Example D.1

Let $\mathbf{A}$ be a 3×3 matrix with elements a_{ij}. Then the determinant of $\mathbf{A}$ is

$$|\mathbf{A}| = a_{11} a_{22} a_{33} + a_{13} a_{21} a_{32} + a_{12} a_{23} a_{31}$$
$$- a_{13} a_{22} a_{31} - a_{12} a_{21} a_{33} - a_{11} a_{23} a_{32}$$

The *cofactor* of the element a_{ij} will be denoted by A_{ij} and is $(-1)^{i+j}$ times the determinant found from $\mathbf{A}$ by omitting the i-th row and j-th row column. It is easy to show that

$$|\mathbf{A}| = \sum_i a_{ji} \, A_{ji} \quad , \quad \text{all } j \qquad \text{(D.1-22)}$$

and that

$$|\mathbf{A}| = \sum_j a_{ji} A_{ji} \quad , \quad \text{all } i \tag{D.1-23}$$

If the determinant of $\mathbf{A}$ is not zero, then its inverse $\mathbf{A}^{-1}$ exists and is given by

$$\mathbf{A}^{-1} = \left[\frac{A_{ij}}{|A|}\right]^T = \frac{A_{ji}}{|A|} \tag{D.1-24}$$

Example D.2

Let $\mathbf{A}$ be the 3×3 matrix given by

$$\mathbf{A} = \begin{bmatrix} 1 & -2 & -1 \\ -1 & 2 & -1 \\ 3 & -2 & 3 \end{bmatrix}$$

Then its determinant $|\mathbf{A}| = 8$ from Example D.1. Its cofactors are $A_{11} = 4$, $A_{12} = 0$, $A_{13} = -4$, $A_{21} = 8$, $A_{22} = 6$, $A_{23} = -4$, $A_{31} = 4$, $A_{32} = 2$, and $A_{33} = 0$. The inverse $\mathbf{A}^{-1}$ is given by Eq. (D.1-24) as

$$\mathbf{A}^{-1} = \begin{bmatrix} \dfrac{4}{8} & 0 & -\dfrac{4}{8} \\[2mm] \dfrac{8}{8} & \dfrac{6}{8} & -\dfrac{4}{8} \\[2mm] \dfrac{4}{8} & \dfrac{2}{8} & 0 \end{bmatrix}^T = \begin{bmatrix} \dfrac{1}{2} & 1 & \dfrac{1}{2} \\[2mm] 0 & \dfrac{3}{4} & \dfrac{1}{4} \\[2mm] -\dfrac{1}{2} & -\dfrac{1}{2} & 0 \end{bmatrix}$$

Quadratic forms - Every $m \times m$ symmetric matrix $\mathbf{A}$ with elements $a_{ij} = a_{ji}$ has associated with it a quadratic form *quadratic form* defined by

$$\mathbf{R}\,\mathbf{A}\,\mathbf{R}^T = \sum_{i=1}^{m} \sum_{j=1}^{m} a_{ij}\, r_i\, r_j \tag{D.1-25}$$

where $\mathbf{R}$ is an arbitrary row vector with elements r_i. The matrix $\mathbf{A}$ is called *positive definite* if

$$\mathbf{R}\,\mathbf{A}\,\mathbf{R}^T > 0 \tag{D.1-26}$$

and is called *non-negative definite* if

$$\mathbf{R} \, \mathbf{A} \, \mathbf{R}^T \geq 0 \qquad \text{(D.1-27)}$$

for all non-zero row vectors $\mathbf{R}$. A positive definite matrix is non-singular as can be seen from Eq. (D.1-20). A matrix will be called *orthogonal* if it has the property that

$$\mathbf{A} \, \mathbf{A}^T = \mathbf{A}^T \, \mathbf{A} = \mathbf{I} \qquad \text{(D.1-28)}$$

D.2 Linear Transformations - Let $\mathbf{X}$ and $\mathbf{Y}$ be $m \times one$ column vectors and let $\mathbf{A}$ be an $m \times m$ matrix. The relationship

$$\mathbf{Y} = \mathbf{A} \, \mathbf{X} \qquad \text{(D.2-1)}$$

is called a *linear transformation* of $\mathbf{X}$ to $\mathbf{Y}$. The transformation from $\mathbf{X}$ to $\mathbf{Y}$ is not necessarily one-to-one. However, if $\mathbf{A}$ is non-singular so that $\mathbf{A}^{-1}$ exists (and is unique), then

$$\mathbf{A}^{-1} \, \mathbf{Y} = \mathbf{A}^{-1} \, \mathbf{A} \, \mathbf{X} = \mathbf{I} \, \mathbf{X} = \mathbf{X} \qquad \text{(D.2-2)}$$

and the inverse (linear) transformation

$$\mathbf{X} = \mathbf{A}^{-1} \, \mathbf{Y} \qquad \text{(D.2-3)}$$

exists proving the one-to-one correspondence. The linear transformation $\mathbf{Y} = \mathbf{B} \, \mathbf{X}$ is orthogonal iff. $\mathbf{B}$ is an orthogonal matrix with the property given by Eq. (D.1-28). Since $|\mathbf{B}|^2 = 1$, such a transformation is nonsingular. If $|\mathbf{B}| = +1$ the transformation corresponds to a *rotation*.

Let $\mathbf{A}$ be an arbitrary nonsingular $n \times n$ matrix and form its *characteristic equation*

$$|\mathbf{A} - \lambda \, \mathbf{I}| = 0 \qquad \text{(D.2-4)}$$

where $\mathbf{I}$ is the identity matrix. This is a polynomial equation in λ of degree n and has n roots $\lambda_1, \lambda_2, ..., \lambda_n$ which are called the *eigenvalues* or *characteristic roots* of the matrix $\mathbf{A}$.

Example D.3

Let $\mathbf{A}$ be a 3×3 matrix given by

$$\mathbf{A} = \begin{bmatrix} 7 & 3 & 0 \\ 3 & 7 & 4 \\ 0 & 4 & 7 \end{bmatrix}$$

The determinant of $\mathbf{A}$ is $|\mathbf{A}| = 168$. The characteristic equation is

$$|A - \lambda I| = \begin{bmatrix} 7-\lambda & 3 & 0 \\ 3 & 7-\lambda & 4 \\ 7 & 4 & 7-\lambda \end{bmatrix}$$

or

$$(\lambda-7)(\lambda-12)(\lambda-2) = 0$$

so that the eigenvalues λ_1, λ_2, and λ_3 are given by

$$\lambda_1 = 7 \; ; \; \lambda_2 = 12 \; ; \; \lambda_3 = 2$$

Note that the eigenvalues $\lambda_1, \lambda_2, ..., \lambda_n$ of an $n \times n$ matrix $\mathbf{A}$ have the properties

$$\sum_{i=1}^{n} \lambda_i \overset{\Delta}{=} \text{trace } \mathbf{A} = \sum_{i=1}^{n} a_{ii} \tag{D.2-5}$$

and

$$\prod_{i=1}^{n} \lambda_i = |\mathbf{A}| \tag{D.2-6}$$

Let $Q(x,x)$ be the quadratic form associated with the matrix $\mathbf{A}$ so that

$$Q(x,x) = \mathbf{X}^T \mathbf{A} \mathbf{X} = \sum_{i=1}^{m} \sum_{j=1}^{n} a_{ij} \, x_i \, x_j \tag{D.2-7}$$

where $\mathbf{X}$ is an arbitrary $n \times one$ column vector. Assume now that $\mathbf{A}$ is *symmetric and positive definite* so that

$$\mathbf{A}^T = \mathbf{A} \tag{D.2-8}$$

and

$$Q(x,x) > 0$$

Now the characteristic equation [Eq. (D.2-4)] has n real roots, and an orthogonal (linear) transformation

$$\mathbf{X} = \mathbf{B}\,\mathbf{Y} \tag{D.2-10}$$

exists which diagonalizes the matrix $\mathbf{A}$; that is, in the y-coordinate system, the matrix $\mathbf{C}$ corresponding to $\mathbf{A}$ is a diagonal matrix (only the elements on the main diagonal differ from zero). We proceed now to find the transformation matrix $\mathbf{B}$ and the form of the diagonal matrix $\mathbf{C}$.

Substitute Eq. (D.2-10) into Eq. (D.2-7) to obtain

$$(\mathbf{B}\,\mathbf{Y})^T \mathbf{A} (\mathbf{B}\mathbf{Y}) = \mathbf{Y}^T (\mathbf{B}^T \mathbf{A} \mathbf{B}) \mathbf{Y} \tag{D.2-11}$$

where $\mathbf{B}^T \mathbf{A} \mathbf{B}$ is required to be a diagonal matrix $\mathbf{C}$ so that

$$\mathbf{Y}^T (\mathbf{B}^T \mathbf{A} \mathbf{B}) \mathbf{Y} = \mathbf{Y}^T \mathbf{C} \mathbf{Y} = \sum_{i=1}^{n} c_{ii} \, y_i^{\,2} \tag{D.2-12}$$

Since the transformation matrix $\mathbf{B}$ is orthogonal, it follows from Eq. (D.1-28) that

$$B^T = B^{-1} \tag{D.2-13}$$

and from Eq. (D.2-12) that

$$B^T \ A \ B = B^{-1} \ A \ B = C \tag{D.2-14}$$

On pre-multiplying this last expression by **B**, we obtain

$$A \ B = B \ C \tag{D.2-15}$$

or

$$\sum_{j=1}^{n} a_{ij} \ b_{jk} = \sum_{j=1}^{n} b_{ij} \ c_{jk} \quad , \quad i,k = 1,2,...,n$$

However, since **C** is diagonal, this equation becomes

$$\sum_{j=1}^{n} a_{ij} \ b_{jk} = b_{ik} \ c_{kk} \quad , \quad i,k = 1,2,...,n$$

Again, this last equation can be written in matrix form as

$$A \ B_k = c_{kk} \ B_k \quad , \quad k=1,2,...,n$$

or

$$(A - c_{kk} \ I) \ B_k = 0 \quad , \quad k=1,2,...,n \tag{D.2-16}$$

where B_k is the column vector

$$B_k = \begin{bmatrix} b_{1k} \\ b_{2k} \\ . \\ . \\ . \\ b_{nk} \end{bmatrix} \tag{D.2-17}$$

Now the original orthogonal transformation matrix **B** is just

$$B = [B_1, B_2, \cdots, B_n] \tag{D.2-18}$$

A comparison of Eqs. (D.2-4) and (D.2-16) shows that a necessary and sufficient condition for Eq. (D.2-16) to have nontrivial solutions for the B_k is that Eq. (D.2-4) hold. Thus $c_{kk} = \lambda_k$ and, for each value of k, the corresponding B_k is found from Eq. (D.2-16). Then the matrix **B** is constructed from Eq. (D.2-17).

Example D.4

Let **A** be the matrix of Example D.3 with eigenvalues $\lambda_1 = 7, \lambda_2 = 12, \lambda_3 = 2$. Thus the diagonal matrix **C** is

$$C = \begin{bmatrix} 7 & 0 & 0 \\ 0 & 12 & 0 \\ 0 & 0 & 2 \end{bmatrix}$$

Note that the determinant of $\mathbf{C}$ is $|\mathbf{C}| = \lambda_1 \lambda_2 \lambda_3 = |\mathbf{A}|$ or, in general, from Eq. (D.2-5),

$$|\mathbf{C}| = \overset{n}{\underset{i=1}{\pi}} \lambda_i = |\mathbf{A}|$$

For $\lambda_1 = c_{11} = 7$; that is $k=1$, Eq. (D.2-16) becomes

$$\begin{bmatrix} 0 & 3 & 0 \\ 3 & 0 & 4 \\ 0 & 4 & 0 \end{bmatrix} \begin{bmatrix} b_{11} \\ b_{21} \\ b_{31} \end{bmatrix} = 0$$

or $3b_{21} = 0$; $3b_{11} + 4b_{31} = 0$; and $4b_{21} = 0$. A solution is the vector of unit length

$$\mathbf{B}_1 = \begin{bmatrix} \dfrac{4}{5} \\ 0 \\ -\dfrac{3}{5} \end{bmatrix}$$

We must now take all $\mathbf{B}_k$ to be of unit length. For $\lambda_2 = c_{22} = 12$, Eq. (D.2-16) is

$$\begin{bmatrix} -5 & 3 & 0 \\ 3 & -5 & 4 \\ 0 & 4 & -5 \end{bmatrix} \begin{bmatrix} b_{12} \\ b_{22} \\ b_{32} \end{bmatrix} = 0$$

or $-5b_{12} + 3b_{22} = 0$; $3b_{12} - 5b_{22} + 4b_{32} = 0$; and $4b_{22} - 5b_{32} = 0$ so that

$$\mathbf{B}_2 = \begin{bmatrix} \dfrac{3}{5\sqrt{2}} \\ \dfrac{1}{\sqrt{2}} \\ \dfrac{4}{5\sqrt{2}} \end{bmatrix}$$

In the same way, for $\lambda_3 = c_{33} = 2$, we obtain

$$\mathbf{B}_3 = \begin{bmatrix} -\dfrac{3}{5\sqrt{2}} \\[2ex] \dfrac{1}{\sqrt{2}} \\[2ex] -\dfrac{4}{5\sqrt{2}} \end{bmatrix}$$

Now the matrix **B** is

$$\mathbf{B} = \begin{bmatrix} \dfrac{4}{5} & \dfrac{3}{5\sqrt{2}} & -\dfrac{3}{5\sqrt{2}} \\[2ex] 0 & \dfrac{1}{\sqrt{2}} & \dfrac{1}{\sqrt{2}} \\[2ex] -\dfrac{3}{5} & \dfrac{4}{5\sqrt{2}} & -\dfrac{4}{5\sqrt{2}} \end{bmatrix}$$

and $\mathbf{X} = \mathbf{B}\,\mathbf{Y}$ becomes the set of equations

$$x_1 = \frac{1}{5\sqrt{2}}\,(4\sqrt{2}\,y_1 + 3y_2 - 3y_3)$$

$$x_2 = \frac{1}{5\sqrt{2}}\,(5y_2 + 5y_3)$$

$$x_3 = \frac{1}{5\sqrt{2}}\,(-3\sqrt{2}\,y_1 + 4y_2 - 4y_3)$$

The quadratic form Q is

$$Q = 7x_1^2 + 6x_1x_2 + 7x_2^2 + 8x_2x_3 + 7x_3^2$$

or

$$Q = 7y_1^2 + 12y_2^2 + 2y_3^2$$

APPENDIX E

THE DIFFERENTIATION OF A DEFINITE INTEGRAL

Let the definite integral $G(u)$ be given by

$$G(u) = \int_{a(u)}^{b(u)} H(x,u)\, dx$$

The derivative of this integral with respect to u is given by

$$\frac{dG}{du} = \int_{a}^{b} \frac{\partial H}{\partial u}\, dx + H(b,u)\frac{db}{du} - H(a,u)\frac{da}{du}$$

if $H(x,u)$ and $\dfrac{\partial H}{\partial u}$ are continuous functions of x and u and if $a(u)$ and $b(u)$ are differentiable functions of u.

APPENDIX F

A MONOTONE NON-DECREASING FUNCTION

In this appendix we prove, as stated in Eq. (8.5-9), that

$$h(p) = \{E(|X|^p)\}^{1/p} \tag{F-1}$$

is a monotone non-decreasing function of p for $p > 0$.

Proof: From Schwarz's inequality, we write

$$\{E(|XY|)\}^2 \le E(|X|^2) E(|Y|^2) \tag{F-2}$$

Let

$$X = |X|^{(p-q)/2}$$

and

$$Y = |X|^{(p+q)/2}$$

for $0 < q < p$. Now Eq. (F-2) becomes

$$\{E(|X|^p)\}^2 \le E(|X|^{p-q}) E(|X|^{p+q})$$

On taking logarithms, we have

$$g(p) = \log\{E(|X|^p)\} \le \frac{1}{2} \log E(|X|^{p-q}) + \frac{1}{2} \log E(|X|^{p+q})$$

or

$$g(p) \le \frac{1}{2} g(p-q) + \frac{1}{2} g(p+q)$$

Suppose the function $g(p)$ exists for some $p > 0$; then it exists for all p_1 such that $0 < p_1 < p$. Furthermore, as $p \to 0$, $g(p) \to 0$. Thus $g(p)$ passes through $(0,0)$ and is convex from below; hence $h(p)$ must be non-decreasing. •

REFERENCES

1. Apostal, T.M. (1974), *Mathematical Analysis,* Addison-Wesley Publishing Co., Inc., Reading, Mass.

2. Birnbaum, Z.W. (1962), *Introduction to Probability and Mathematical Statistics,* Harper and Brothers, New York, N.Y.

3. Breiman, L. (1968), *Probability,* Addison-Wesley Publishing Co., Inc., Reading, Mass.

4. Burington, R.S. (1973), *Handbook of Mathematical Tables and Formulas,* Handbook Publishers, Inc., Sandusky, O.

5. Campbell, G.A., and R.M. Foster, (1948), *Fourier Integrals for Practical Applications,* D. Van Nostrand Company, Inc., Princeton, N.J.

6. Courant, R. (1937), *Differential and Integral Calculus,* Vols. 1 and 2, Interscience Publishers, Inc., New York, N.Y.

7. Cramer, H. (1946), *Mathematical Methods of Statistics,* Princeton University Press, Princeton, N.J.

8. Dirac, P.A.M. (1958), *Principles of Quantum Mechanics,* Oxford University Press, Oxford.

9. Doob, J.L. (1953), *Stochastic Processes,* John Wiley & Sons, Inc., New York, N.Y.

10. Dubes,R.C. (1968), *The Theory of Applied Probability,* Prentice-Hall, Inc., Englewood Cliffs, N.J.

11. Feller, W. (1968), *An Introduction to Probability Theory and Its Applications,* Vol. I, John Wiley & Sons, Inc., New York, N.Y.

12. Feller, W. (1966), *An Introduction to Probability Theory and Its Applications,* Vol. II, John Wiley & Sons, Inc., New York, N.Y.

13. Hille, Einar (1979), *Analysis,* Vols. 1 and 2, Blaisdell Publishing Company, New York, N.Y.

14. Hohn, Franz (1973), *Elementary Matrix Algebra,* The Macmillan Company, New York, N.Y.

15. Jolley, L.B.W. (1961), *Summation of Series,* Dover Publications, Inc., New York, N.Y.

16. Kendall, M.G. and A. Stuart, (1977), *The Advanced Theory of Statistics,* Vol. I, Hofner Publishing Company, New York, N.Y.

17. Kolmogorov, A.N. (1956), *Foundations of the Theory of Probability,* Chelsea Publishing Company, New York, N.Y.

18. Lighthill, M.J. (1958), *An Introduction to Fourier Analysis and Generalized Functions,* Cambridge University Press, Cambridge.

19. Munroe, M.E. (1953), *Introduction to Measure and Integration,* Addison-Wesley Publishing Co., Inc., Reading, Mass.

20. Neveu, J. (1965), *Mathematical Foundations of the Calculus of Probability,* Holden-Day, Inc., San Francisco, Calif.

Table I - The Normal Distribution

x	f(x)	F(x)	x	f(x)	F(x)
0.00	0.3989	0.5000	1.40	0.1497	0.9192
0.05	0.3984	0.5199	1.50	0.1295	0.9332
0.10	0.3970	0.5398	1.60	0.1109	0.9452
0.15	0.3945	0.5596	1.70	0.0940	0.9554
0.20	0.3910	0.5793	1.80	0.0790	0.9641
0.25	0.3867	0.5987	1.90	0.0656	0.9713
0.30	0.3814	0.6179	2.00	0.0540	0.9772
0.35	0.3752	0.6368	2.10	0.0440	0.9821
0.40	0.3683	0.6554	2.20	0.0355	0.9861
0.45	0.3605	0.6736	2.30	0.0283	0.9893
0.50	0.3521	0.6915	2.40	0.0224	0.9918
0.55	0.3429	0.7088	2.50	0.1075	0.9938
0.60	0.3332	0.7257	2.60	0.0136	0.9953
0.65	0.3230	0.7422	2.70	0.0104	0.9965
0.70	0.3123	0.7580	2.80	0.00791	0.99744
0.75	0.3011	0.7734	2.90	0.00595	0.99813
0.80	0.2897	0.7881	3.00	0.00443	0.99865
0.85	0.2780	0.8023	3.20	0.00238	0.99931
0.90	0.2661	0.8159	3.40	0.00123	0.99966
0.95	0.2541	0.8289	3.60	0.00061	0.99984
1.00	0.2420	0.8413	3.80	0.00029	0.99993
1.10	0.2179	0.8643	4.00	0.00013	0.99997
1.20	0.1942	0.8849	4.50	0.000016	0.999996
1.30	0.1714	0.9032	5.00	0.0000015	0.9999997

$$f(x) = \frac{1}{\sqrt{2\pi}}\, e^{-\frac{x^2}{2}} \qquad F(x) = \frac{1}{\sqrt{2\pi}} \int_{-\infty}^{x} e^{-\frac{y^2}{2}}\, dy$$

INDEX

SPRINGER TEXTS IN ELECTRICAL ENGINEERING